Master Big Bangs through Black Holes in Four Hours

An Integrated Theory of Everything Introduction

Antonio A. Colella

Copyright © September 2012 by Antonio A. Colella
All rights reserved.
ISBN: 147931093X
ISBN 13: 9781479310937
Library of Congress Control Number: 2012917267
CreateSpace Independent Publishing Platform
North Charleston, South Carolina

This book is dedicated to my mother, Angela;
my father, Berardino; my sister, Patricia;
and my daughters, Stacy and Whitney.

Contents

Preface   ix

Chapter 1
Introduction   1

Chapter 2
Five Matter Particles   5

Chapter 3
Two Dark Matter Particles   9

Chapter 4
Exponential Notation   13

Chapter 5
Six Force Particles   19

Chapter 6
Super and Higgs Forces   25

Chapter 7
Dark Energy   31

Chapter 8
String Theory   35

Chapter 9
Universe Expansions 43

Chapter 10
Big Bang 47

Chapter 11
Our Universe 53

Chapter 12
Stellar Black Holes 61

Chapter 13
Super Supermassive Quark Star (Matter)/Black Hole (Energy) 69

Chapter 14
Super Universe and Our Precursor Universe 77

Chapter 15
Cosmological Constant Problem 81

Chapter 16
Black Hole Information Paradox 89

Chapter 17
Matter/Antimatter 93

Chapter 18
Quantum Gravity Theory 99

Chapter 19
Systems Engineering 101

Chapter 20
Conclusions 109

Endnotes 111

Glossary 127

Further Readings 147

Appendix A An Integrated Theory of Everything Research Article 149

Research Article Endnotes 199

Appendix B An Integrated Theory of Everything Video Presentation 209

Preface

Master Big Bangs through Black Holes in Four Hours is the popular title for *An Integrated Theory of Everything Introduction.* The popular title's significance is that the Super Universe, which consists of parallel universes including our own, was created by cycles of big bangs through stellar black holes.

The Theory of Everything (TOE) has been described as the final theory, the crowning achievement of science, the ultimate triumph of human reasoning, and the knowledge of God's mind. After almost a century of research by physicists including Albert Einstein and Stephen Hawking, it remains the foremost unsolved problem of physics.

An Integrated TOE unifies all known physical phenomena from the infinitely small or Planck cube scale to the infinitely large or Super Universe scale. Each fundamental matter and force particle exists as a string in a Planck cube. Any object in the Super Universe can be represented by a volume of contiguous Planck cubes containing fundamental matter or force particle strings.

The foundations of an Integrated TOE are twenty independent existing theories, including string, matter and force particle creation, Higgs forces (Higgs bosons or God particles), dark matter, universe expansions, dark energy, Super Universe, stellar black holes, cosmological constant problem, black hole information paradox, and baryogenesis (prominence of matter over anti-matter). The premise of an Integrated TOE is without sacrificing their integrities; these twenty independent existing theories are replaced by twenty interrelated amplified theories. If each independent existing theory is represented by a jigsaw puzzle piece, they do not fit together because of their independence. Selectively adding new interrelated requirements to the twenty independent existing theories is equivalent to selectively amplifying the sizes and shapes of twenty jigsaw puzzle pieces. This selective amplification process produces twenty snuggly fitting interrelated and amplified jigsaw puzzle pieces for an Integrated TOE.

Amplification of Higgs force theory is essential for an Integrated TOE. Amplifications include super force condensations or creation of particles occurs for five matter particles and their five associated Higgs forces; matter particles and their associated Higgs forces are one and inseparable; and the sum of five permanent Higgs force energies is dark energy.

This Integrated TOE introduction book is intended for a broad spectrum of readers. The book is readable by students with a basic understanding of physics or chemistry without referring to book endnotes. Reading this book and its endnotes by students, scientists, engineers, and mathematicians provides a bridge to an Integrated TOE research article in appendix A. The latter is intended for PhD researchers in particle physics and cosmology.

Chapter 1

Introduction

An Integrated TOE unifies all known physical phenomena, from the infinitely small or Planck cube scale to the infinitely large or Super Universe scale.[1]

A million is a one followed by six zeros. A billion is a one followed by nine zeros. A trillion is a one followed by twelve zeros. A googol is a one followed by one hundred zeros. A Planck cube is infinitely small because almost a googol of Planck cubes fits in each grain of sand. Our universe consists of one hundred billion galaxies, each containing one hundred billion stars. The Super Universe is infinitely large because the Super Universe contains over a googol of parallel universes.

Each of eleven fundamental matter and force particle listed in table 1 exists within a Planck cube, the universe's fundamental building block. The Planck cube is the quantum or unit of matter particle, force particle, and space. Any Super Universe object is representable by a volume of contiguous Planck cubes containing matter or force particles. Planck cubes are visualized as infinitely small, cubic, Lego blocks. A volume of contiguous Planck cubes can represent a proton, an atom, a molecule, an encyclopedia, a person, Mount Everest, a star, a galaxy, our universe, or the entire Super Universe. Figure 1 shows an encyclopedia consisting of a volume of contiguous Planck cubes. The figure is not to scale because there are over a googol of Planck cubes in an encyclopedia. An encyclopedia is composed of ink, paper, and binder molecules. Molecules can be decomposed into atoms. Atoms can be decomposed into protons, neutrons, and electrons. Protons and neutrons can be decomposed into fundamental matter particles, up quarks and down quarks. Every fundamental matter particle, such as an up quark in each of the encyclopedia's molecules, exists within a separate Planck cube.

The Super Universe consists of nested universes. Figure 2, Super Universe, shows our universe nested within our precursor universe, which is nested within the Super Universe. Our universe, our precursor universe, and the Super Universe are spherical volumes. Nested within our universe are our Milky Way galaxy and approximately one hundred billion other galaxies. The Super Universe was created first before our precursor universe, and our precursor universe was created before our universe. Figure 2 is not to scale because the Super Universe is more than a googol times larger than our universe.

This book is readable without referring to the book endnotes. However, the endnotes provide a bridge to an Integrated TOE research article in appendix A.

Particle	Matter	Force
up quark	x	
down quark	x	
electron	x	
zino	x	
photino	x	
gravitational		x
electromagnetic		x
strong		x
weak		x
super (mother)		x
Higgs		x

Table 1. Fundamental matter and force particles.

Introduction

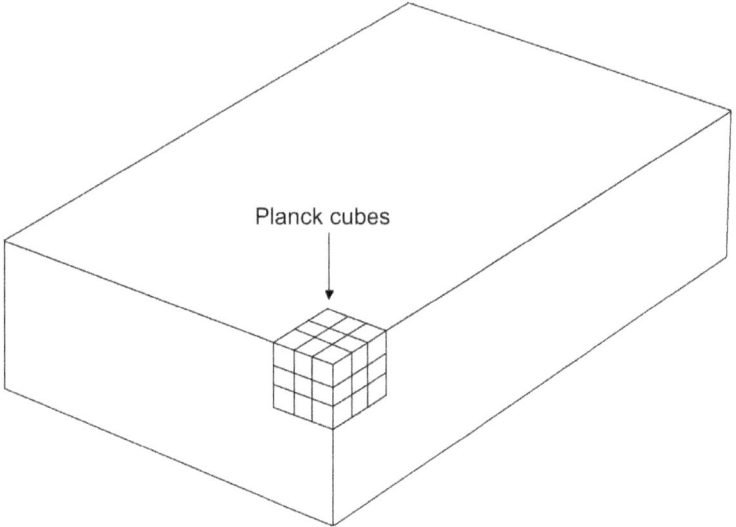

Figure 1. Encyclopedia's Planck cubes.

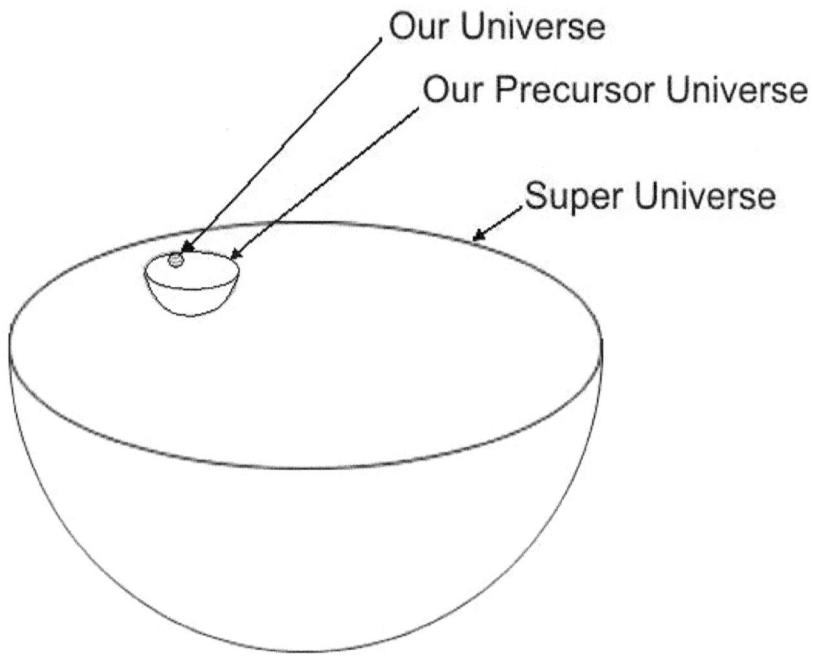

Figure 2. Super Universe.

Chapter 2

Five Matter Particles

Molecules can be decomposed into atoms. Atoms can be decomposed into protons, neutrons, and electrons. Protons and neutrons can be decomposed into fundamental matter particles, up quarks, and down quarks. Five fundamental matter particles are shown in table 1; three are known atomic matter particles (up quark, down quark, and electron) and two are proposed but undetected dark matter particles (zino and photino).[2] Atoms are described by the periodic table of elements.

Molecular matter has three forms: solid like a mountain, liquid like water, or gas like the atmosphere. With their acceptance of molecules in the early nineteenth century, scientists defined molecules as the basic building blocks of matter. Any solid, liquid, or gas could be decomposed into molecules such as water (H_2O) and carbon dioxide (CO_2). With the publication of Dmitri Mendeleev's first periodic table in the midnineteenth century, molecules could be decomposed into the second atomic matter building block level, atoms. A water molecule could be decomposed into two hydrogen and one oxygen atoms and the carbon dioxide molecule into one carbon and two oxygen atoms.

Following Ernest Rutherford's discovery of an atom's nucleus in the early twentieth century, atoms could be decomposed into the third atomic matter building block level, protons and neutrons[3] in the atom's nucleus or center and electrons in shells outside the nucleus. For example, the hydrogen atom has one proton in its nucleus and one electron in its shell, and the helium atom has two protons and two neutrons in its nucleus and two electrons in its shells. Protons have a positive charge; electrons have a negative charge, whereas neutrons have no charge. Atomic number is the number of protons in the atom's nucleus, which identifies a chemical element. Mass number is the number of protons and neutrons in the nucleus. The number of neutrons in the nucleus is the mass number minus the atomic number. Since the number of neutrons in the nucleus determines isotopes of a chemical element, isotopes have the same atomic number but different mass numbers. For example, carbon (atomic number 6), with a mass number of 12, has six protons and six neutrons, whereas carbon, with a mass number 13, has six protons and seven neutrons.

The above were the three identified atomic matter building block levels before my departure from academia in 1961. During my forty-four-year industrial career, my tasks included specifying system requirements and systems integration/tests for complex electronic projects such as air traffic control systems; a weather and radar processor; FBI Integrated Automated Fingerprint Identification System; a global positioning system (GPS)[4]; nuclear submarine sonars and trainers; helicopter flight, tactical, and maintenance trainers; a manned orbiting laboratory and simulator; a worldwide military command and control system; missile defense systems; and phased array radars.

Upon my return to academic studies in 2006, and to my surprise, scientists had decomposed matter into two additional atomic matter building block levels. In 1964, Murray Gell-Mann and George Zweig independently proposed the fourth atomic matter building block level, which decomposed protons and neutrons into fundamental matter particles, up, and down quarks. The proton could be decomposed into two up quarks and one down quark and the neutron into one up quark and two down quarks. The three fundamental atomic matter particles (positive electrically charged up quark, negative electrically charged down quark, and negative electrically charged electron) could be further decomposed into strings or the fifth atomic matter building block level, as described in chapter 8, "String Theory."

Table 1 shows five fundamental matter particles: three known atomic matter particles (up quark, down quark, and electron) and two proposed but undetected dark matter particles (zino and photino). The three fundamental atomic matter particles, (up quark, down quark, and electron) exist within their own Planck cube and are the building blocks of all atomic matter. For example, a volume of contiguous Planck cubes can represent a proton, an atom, a molecule, an encyclopedia, a person[5], Mount Everest, a star, a galaxy, our universe, or the entire Super Universe.

Atoms are described by the periodic table of elements, shown in figure 3, and include hydrogen (H), helium (He), oxygen (O), nitrogen (N), iron (Fe), silver (Ag), gold (Au), lead (Pb), etc. The other two fundamental dark matter particles (zino and photino) are nonatomic and are described in chapter 3, "Two Dark Matter Particles."

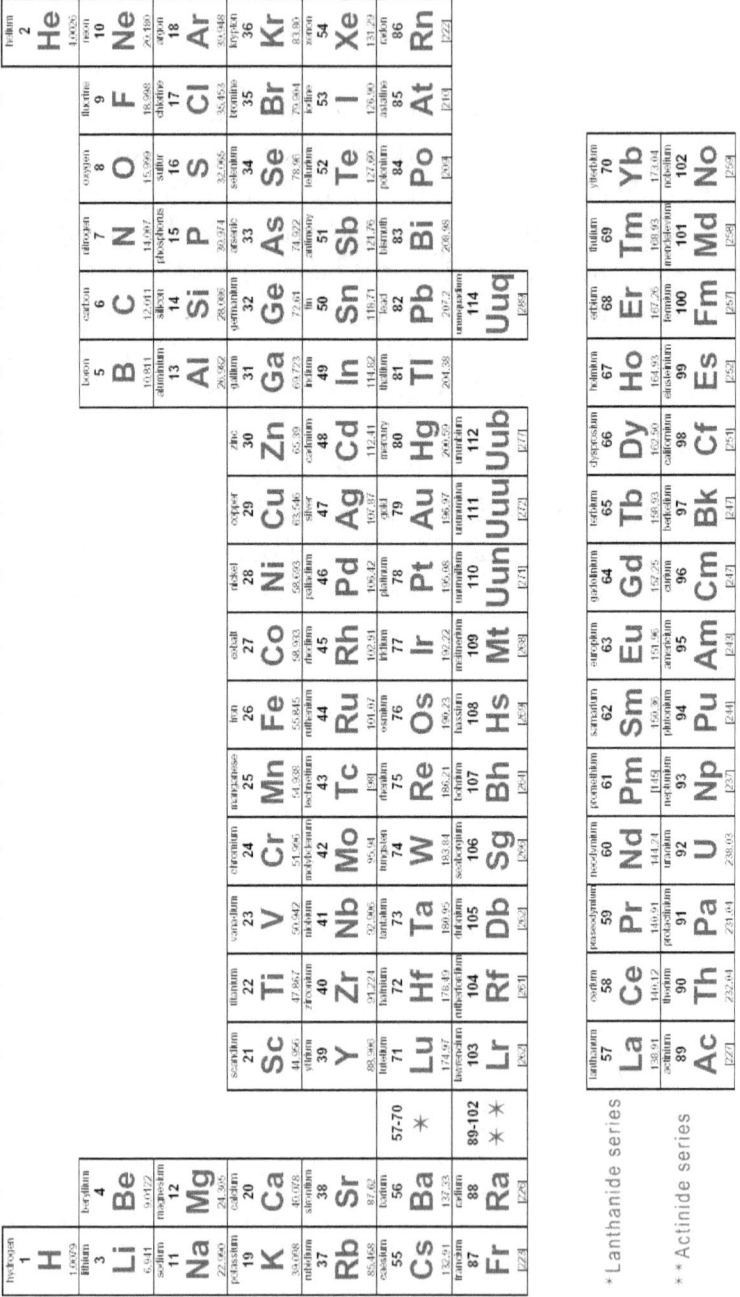

Figure 3. Periodic table of elements.

Chapter 3

Two Dark Matter Particles

The two undetected fundamental dark matter particles (zino and photino) were nonatomic. Dark matter was originally proposed to account for our universe's missing mass. Matter creation occurred immediately following the start of our universe and consisted of atomic matter (up quark, down quark, and electron) and dark matter (zino and photino) fundamental particles. At 30,000 years after the start of our universe, dark matter began to clump together to form the framework of our universe's galaxies and begin the galaxy evolution process.

The two undetected fundamental dark matter particles (zino and photino) are nonatomic and exist within their own Planck cubes. Conceptually, a second currently undefined periodic table of dark elements exists which defines nonatomic dark matter, just as the figure 3 periodic table of elements defines atomic matter, such as hydrogen, helium, oxygen, and carbon.[6]

Dark matter was originally proposed to account for our universe's missing mass. The missing mass was necessary to explain the rotational velocity of stars in spiral galaxies, which acted as if the galaxies contained mass in addition to known atomic mass. Additional astronomical observation evidence corroborated dark matter existence.[7] Atomic and dark matter mass comprised 27.4% of our universe's total energy/mass. Dark matter accounted for 22.8% and atomic matter 4.6%.[8] Thus, the missing dark matter mass was approximately five times that of atomic mass. Energy/mass was the equivalency of energy and mass described by Einstein's equation $E = mc^2$, where E is energy, m is mass, and c is the velocity of light. Matter particles were described by energy/mass, whereas force or energy particles were described by energy.

Three dark matter detection techniques are currently being internationally investigated; direct, indirect, and collider. In the direct technique, rare dark matter collisions with atoms are detected. In the indirect technique, the annihilation products of dark matter are detected. In the collider technique, dark matter produced by particle accelerators is detected by particle accelerator detectors. In spite of these sophisticated detection techniques, dark matter has still not been detected.[9]

Matter creation occurred immediately following the start of our universe and consisted of atomic matter (up quark, down quark, and electron) and dark matter (zino and photino) fundamental particles. Our universe had uniform distribution of matter from matter creation to 30,000 years after the start of our universe. Uniform distribution meant samples of our universe before 30,000 years contained the same number of atomic matter (up quark, down quark, and electron) and dark matter (zino and photino) particles.

At 30,000 years after the start of our universe, dark matter began to clump together to form the framework of our universe's galaxies and begin the galaxy evolution process. Dark matter clumping was also the beginning of our universe's nonuniform distribution of matter. Nonuniform distribution of matter after 30,000 years meant samples of our universe did not contain the same number of atomic matter (up quark, down quark, and electron) and dark matter (zino and photino) particles.

Between the 30,000 and 380,000 years transition period, dark matter clumped together, whereas electrically charged atomic matter, such as positively charged protons (hydrogen nuclei), positively charged helium nuclei, and negatively charged electrons, did not. That is, between 30,000 and 380,000 years, atomic matter (up quarks, down quarks, and electrons) was uniformly distributed, whereas dark matter (zinos and photinos) was not. At 380,000 years after the start of our universe, our universe cooled sufficiently for the negatively charged electrons to combine with positively charged protons and helium nuclei.[10] This nuclei and electron combining process produced electrically neutral hydrogen and helium atoms, which began clumping around the dark matter framework.

Chapter 4

Exponential Notation

Exponential notation is a method of describing the extreme range of numbers required for the infinitely large and small numbers in an Integrated TOE. Exponential notation eliminates zeros and simplifies multiplication and division of infinitely large and small numbers. Since the Planck length is 1.6 x 10^{-35} meters, there is almost a googol of Planck cubes in each grain of sand.

Exponential notation is a method of describing the infinitely large numbers in an Integrated TOE. A thousand is a one followed by three zeros. A million is a one followed by six zeros. A billion is a one followed by nine zeros. A trillion is a one followed by twelve zeros. A googol is a one followed by one hundred zeros. In exponential notation, a thousand or 1,000 is expressed as 1.0 x 10^3 with the exponent 3 signifying the decimal point in 1.0 moves three digits to the right to reconstitute the number 1,000. A million or 1,000,000 is expressed as 1.0 x 10^6 with the exponent 6 signifying the decimal point in 1.0 moves six digits to the right to reconstitute the number 1,000,000. A billion is expressed as 1.0 x 10^9, a trillion as 1.0 x 10^{12}, and a googol as 1.0 x 10^{100}. With the 1.0 understood, a thousand is implicitly expressed as 10^3, a million as 10^6, a billion as 10^9, a trillion as 10^{12}, and a googol as 10^{100}. Examples of numbers other than one are four thousand or 4,000, which is explicitly expressed as 4.0 x 10^3, and seven billion or 7,000,000,000, which is explicitly expressed as 7.0 x 10^9.

Exponential notation is a method of describing the infinitely small numbers in an Integrated TOE. For fractions in exponential notation, a thousandth (.001) is expressed as 1.0 x 10^{-3} with the exponent -3 signifying the decimal point in 1.0 moves three digits to the left to reconstitute the number .001. One millionth (.000001) is expressed as 1.0 x 10^{-6} with the exponent -6 signifying the decimal point in 1.0 moves six digits to the left to reconstitute the number .000001. A billionth (.000000001) is expressed as 1.0 x 10^{-9}, a trillionth as 1.0 x10^{-12}, and a googolth[11] as 1.0 x 10^{-100}. With the 1.0 understood, a thousandth is implicitly expressed as 10^{-3}, a millionth as 10^{-6}, a billionth as 10^{-9}, a trillionth as 10^{-12}, and a googolth as 10^{-100}. An example of a number other than one is five millionths or .000005, which is explicitly expressed as 5.0 x 10^{-6}.

Exponential notation eliminates zeros and simplifies multiplication and division of infinitely large numbers. Mathematics is simplified because the exponents are added for multiplication (10^a x 10^b = 10^{a+b}) and subtracted for division ($10^a/10^b$ = 10^{a-b}). One million (1,000,000) multiplied by one thousand (1,000) equals one billion (1,000,000,000). In exponential notation, one million (10^6) multiplied by one thousand (10^3) equals one billion, or 10^6 x 10^3 = 10^{6+3} = 10^9. One billion (1,000,000,000) divided by one million (1,000,000) equals one thousand (1,000). In exponential notation, one billion (10^9) divided by one million (10^6) equals one thousand or $10^9/10^6$ = 10^{9-6} = 10^3.

Exponential notation eliminates zeros and simplifies multiplication and division of infinitely small numbers. The formulas for multiplication and division of fractions are identical to those for integer numbers. For example, a millionth (.000001) multiplied by a thousandth (.001) equals one billionth (.000000001). In exponential notation, a millionth (10^{-6}) multiplied by a thousandth (10^{-3}) equals one billionth or 10^{-6} x 10^{-3} = $10^{-6+(-3)}$ = 10^{-9}. A thousandth (.001) divided by a millionth (.000001) equals one thousand (1000). In exponential notation, a thousandth (10^{-3}) divided by a millionth (10^{-6}) equals $10^{-3}/10^{-6} = 10^{-3-(-6)} = 10^3$.

Since the Planck length is 1.6×10^{-35} meters, there is almost a googol of Planck cubes in each grain of sand. Figure 4 shows a greatly magnified Planck cube with a Planck length edge of .00000000000000000000000000000000016 meters, or 1.6×10^{-35} meters in exponential notation. The Planck cube volume equals Planck length cubed or $(1.6 \times 10^{-35})(1.6 \times 10^{-35})(1.6 \times 10^{-35}) = 4.1 \times 10^{-105}$ cubic meters. The volume of a grain of sand is its length, 1×10^{-3} meters, cubed, or 1×10^{-9} cubic meters. The number of Planck cubes in a grain of sand is the volume of the grain of sand divided by the Planck cube volume, or $(1 \times 10^{-9}$ cubic meters$)/(4.1 \times 10^{-105}$ cubic meters$) = .24 \times 10^{96}$. Thus, there is almost a googol (10^{100}) of Planck cubes in each grain of sand.

The above examples are summarized in table 2, "Numbers in decimal and exponential notations."

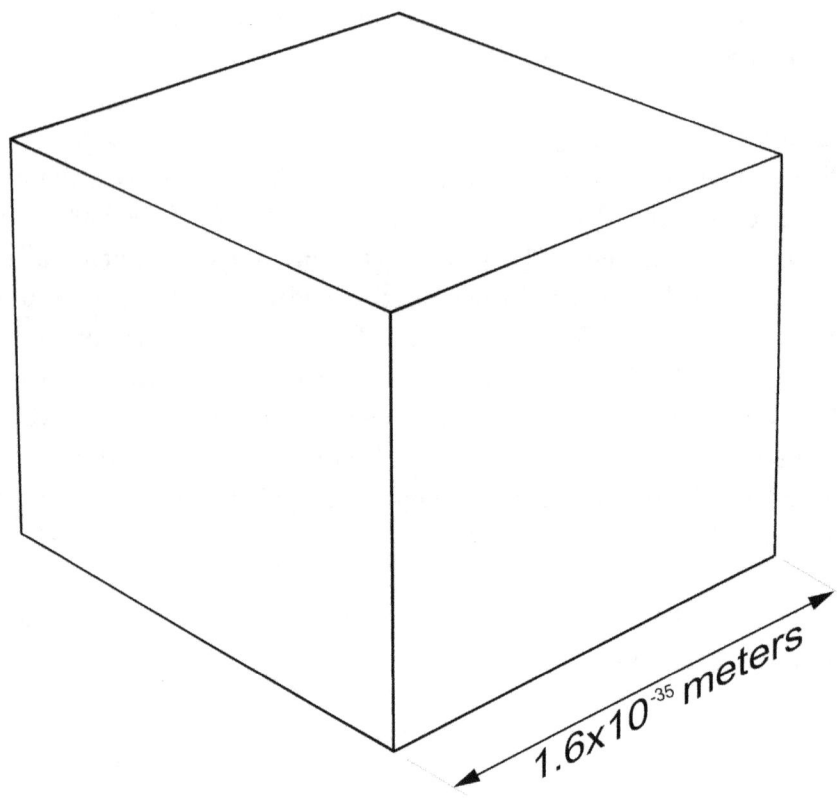

Figure 4. Planck cube.

Exponential Notation

Number	Decimal Notation	Exponential Notation
one	1	10^0
ten	10	10^1
hundred	100	10^2
thousand	1,000	10^3
million	1,000,000	10^6
billion	1,000,000,000	10^9
trillion	1,000,000,000,000	10^{12}
googol	1 followed by one hundred zeros	10^{100}
tenth	.1	10^{-1}
hundredth	.01	10^{-2}
thousandth	.001	10^{-3}
millionth	.000001	10^{-6}
billionth	.000000001	10^{-9}
trillionth	.000000000001	10^{-12}
googolth	decimal point followed by 99 zeros and a one	10^{-100}
four thousand	4,000	4.0×10^3
seven billion	7,000,000,000	7.0×10^9
five millionths	.000005	5.0×10^{-6}

Table 2. Numbers in decimal and exponential notations.

Chapter 5

Six Force Particles

Six fundamental force or energy particles are shown in table 1; four known (gravitational, electromagnetic, strong, and weak) and two proposed but undetected (super and Higgs). All six exist within their own Planck cube. Atom production occurs during three astronomical processes. The strength of the gravitational force relative to the electromagnetic force is 10^{-42} and is known as the hierarchy problem.

The gravitational force is the attractive force between masses, such as the earth holding people on its surface and the moon in its orbit. The electromagnetic force is the attractive force of two opposite polarity electromagnets and the repulsive force of two like polarity electromagnets. The electromagnetic force holds electrons and protons together inside atoms and electrons of atoms together inside molecules. The strong force is the nuclear binding force between up quarks and down quarks within a proton or neutron. The strong force also binds protons and neutrons together in a nucleus.[12] An example of the weak force is the transformation of down quarks into up quarks and vice versa.

In the late sixteenth century, Galileo Galilei demonstrated the constant gravitational force acceleration of all objects. In the late seventeenth century, Sir Isaac Newton developed the gravitational force law. Newton's law stated gravitational force was proportional to the product of two masses and inversely proportional to the square of the range between the masses. Newton's equation was $F = Gm_1m_2/r^2$, where m_1 and m_2 were two masses, r was the range between masses, and G was the gravitational constant.[13]

Coulomb's electrostatic law between two charged particles at rest has the same form as Newton's law, and its equation is $F = Cq_1q_2/r^2$, where q_1 and q_2 are two charges, r is the range between charges, and C is Coulomb's constant. If the charges are moving, they experience a combined electrostatic and magnetic force called the electromagnetic force. A changing electric field generates a magnetic field and vice versa. Electricity and magnetism were originally thought of as two separate forces, but in the late nineteenth century, James Maxwell unified the two. Electromagnetic radiation types consist of gamma rays, x-rays, ultraviolet rays, visible light rays, infrared rays, radar, frequency modulated (FM) radio waves, television (TV) waves, shortwave radio waves, and amplitude modulated (AM) radio waves, as shown in figure 5, "Electromagnetic spectrum."[14] Electromagnetic radiation types differ in their frequency and energy, where energy is proportional to frequency. From figure 5, gamma rays have the shortest wavelength, highest frequency, and highest energy.[15] In contrast, amplitude modulated (AM) radio waves have the longest wavelength, lowest frequency, and lowest energy. All electromagnetic radiation types exist around us, but only light rays are visible to our eyes, or detectable by us. However, if an electromagnetic radiation detector (e.g., an AM radio or TV) is brought in our vicinity, those types of electromagnetic radiation are detectable. Similarly, all other electromagnetic radiation types, such as gamma rays, x-rays, ultraviolet rays, infrared rays, radar, frequency modulated (FM) radio waves, and shortwave radio waves, are detectable with the appropriate electromagnetic radiation detector.

The strong force between two quarks depends on whether the quark separation is less or greater than a proton radius.[16] The equation for the strong force is $F = \alpha_s/r^2 - k$, where r is quark separation and α_s and k are constants.[17] Thus, the strong force equation has two components: a Newton-like force which is inversely proportional to the square of the range between quarks, and a constant force. For quark separations less than a proton radius, the strong force is defined by the Newton-like term. If the separation between quarks is greater than a proton radius, the force is a constant. In the latter case, fields form which attract the quarks as if they are confined by an elastic bag.

In 1968, Sheldon Glashow, Abdus Salam, and Steven Weinberg developed the electroweak theory, which defined the electromagnetic force and weak force to be components of the combined electroweak force. The weak force transformed a down quark into an up quark.[18] Similarly, with the addition of energy, the weak force transformed an up quark into a down quark.[19] Since a neutron consists of one up quark and two down quarks and a proton consists of two up quarks and one down quark, if a down quark in a neutron is transformed into an up quark, the neutron is transformed into a proton. Since the number of protons in the atom's nucleus identifies a chemical element, transforming a neutron into a proton transforms an atom into a different atom type. Atom production occurs during three astronomical processes: big bang helium nuclei formation, formation of large stellar cores, and supernova explosions. The first process is described in chapter 10, "Big Bang," and the second and third processes are described in chapter 11, "Our Universe." Because of its ability to transform an atom into a different type of atom, the weak force is the alchemist of our universe.

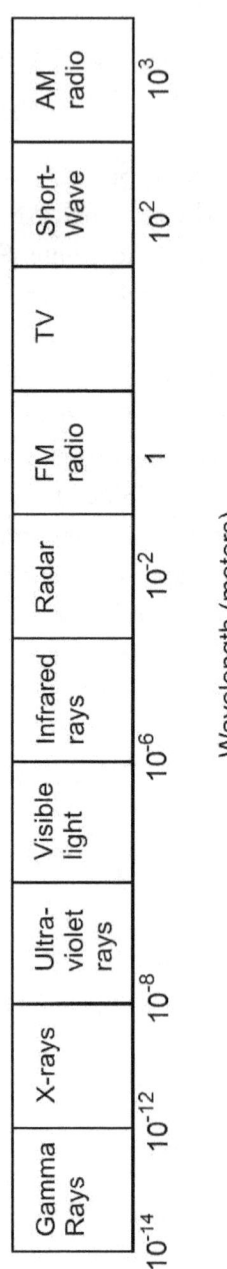

Figure 5. Electromagnetic spectrum.

The strength of the gravitational force relative to the electromagnetic force is 10^{-42} and is known as the hierarchy problem.[20] Of the four known forces, the gravitational force is counterintuitively the weakest, whereas the strong force is the strongest. Weakness of the gravitational force is demonstrated by a paper clip and a small electromagnet. The entire mass of the earth holds the paper clip on a table, whereas a small electromagnet counteracts all the earth's gravity by lifting the paper clip from the table. The weakness of the gravitational force relative to the electromagnetic force is a factor of 10^{-42}. For distances much less than a proton's radius, the strengths of the electromagnetic and weak forces are equal. The electromagnetic/weak forces are 10^{-2} the value of the strong force. Table 3 summarizes the strengths of forces relative to the strong force.

Force	Relative Strength
Strong	1
Electromagnetic/Weak	10^{-2}
Gravitational	10^{-44}

Table 3. Relative strengths of forces.

Chapter 6

Super and Higgs Forces

Two proposed but undetected force particles were the super and Higgs forces (Higgs bosons).[21] The super force was the mother of all matter and force particles proposed by physicists in the unified TOE. Amplifications of Higgs forces theory for an Integrated TOE included: super force condensations occurred for five matter particles and their five associated Higgs forces during particle creation; Higgs forces were residual super forces containing characteristics (e.g., mass, charge, spin) of their associated matter particles; mass was given to a matter particle via its associated Higgs force and the gravitational force messenger particles transmitted between matter particles; matter particles and their associated Higgs forces were one and inseparable; the Higgs mechanism was bidirectional, supporting condensations from and evaporations to the super force; and the sum of five permanent Higgs force energies was dark energy.

Peter Higgs proposed the Higgs force; Higgs force properties, and the Higgs mechanism[22] between 1964 and 1966. An Integrated TOE was developed by selectively amplifying independent existing theories to interrelated amplified theories as described in chapter 19, "Systems Engineering." Amplification of Higgs force theory was essential to development of an Integrated TOE.

The super force was the mother of all matter and force particles proposed by physicists in the unified TOE. At the extremely high temperatures at the start of our universe, only super force particles existed and were the mother particles of all five fundamental matter particles (up quark, down quark, electron, zino, and photino) and all five fundamental force particles (gravitational, electromagnetic, strong, weak, and Higgs). Thus, at the start of our universe, all matter and force particles were unified as super force particles.

Super force condensations occurred for five matter particles and their five associated Higgs forces during particle creation. At the start of our universe, the volume of our universe expanded and our universe's temperature cooled. Following particle creation, each of five matter and four force particles had an associated Higgs particle. Similar to the three condensation phases from steam to water to ice, a super force particle condensed into a matter or force particle and its associated Higgs particle at a specific temperature. For example, a super force particle condensed or froze out to an up quark matter particle and its associated Higgs force. During particle creation, condensations occurred at nine different temperatures for five matter and four force particles, with their nine associated Higgs particles. However, for simplicity, in this book only five matter particles and their five associated Higgs forces are analyzed.[23]

The process of generating five matter particles (up quark, down quark, electron, zino, and photino) and their five associated Higgs force particles is the Higgs mechanism. The up quark Higgs mechanism, which resembles a rolling ball on a Mexican sombrero, is shown in figure 6. The Z axis represents energy density or energy per cubic meter. The amount of energy available for condensation is energy density multiplied by our universe's volume at the condensation time, that is, Energy = (Energy/unit volume) (volume). The X axis represents one Higgs force particle's energy associated with an up quark matter particle.

A particle can be in either of two states, super force or a condensed matter particle and its associated Higgs force. Each of the two states corresponds to a unique ball position on the Mexican sombrero. When the ball is at its peak position at the top of the sombrero at x = 0, z = 2 of figure 6, only super force particles exist because condensation to up quarks and their associated Higgs forces has not yet occurred. When the ball rolls down to the x = -2, z = 1.5 position shown in figure 6, super force particles have condensed to up quarks and their associated Higgs forces. The Higgs force particle's energy associated with one up quark is the figure 6 ball X position, that is, x = -2. The energy density on the Z axis between 1.5 and 2.0 condensed into up quarks, and the remaining energy density between 0.0 and 1.5 condensed into their associated Higgs force particles.

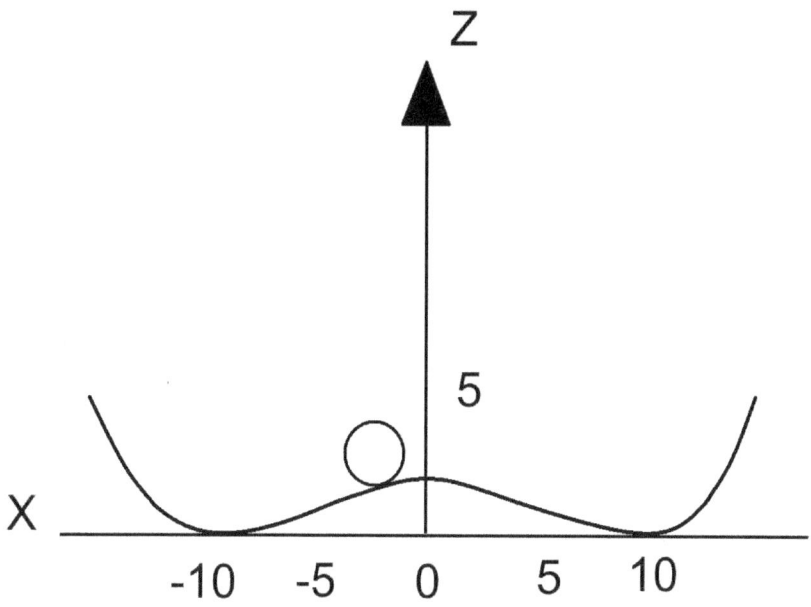

Figure 6. Up quark Higgs mechanism.

There were five unique Higgs mechanisms for the five matter particles, each having the same generic shape of figure 6. Each matter particle had a different value for the initial super force energy density and the matter particle's associated Higgs force particle's energy. That is, each of the five Higgs mechanisms had a unique peak position on the Z axis and a unique figure 6 ball X position. Also, each of the five matter particle condensations occurred at a different time or temperature during matter creation, with the heaviest matter particle (either the zino or photino) condensing first and the lightest matter particle (electron) condensing last.

Higgs forces are residual super forces containing characteristics (e.g., mass, charge, spin) of their associated matter particles. The Higgs force is visualized as a three-dimensional field surrounding and attached to its associated matter particle or symbolically as a single Planck cube containing the Higgs force and attached to its associated matter particle.

Mass is given to a matter particle via its associated Higgs force and the gravitational force messenger particles transmitted between matter particles. The gravitational force messenger particle contains an embedded computer and clock. For example, an up quark's mass (m_1) is extracted from an up quark's associated Higgs force by the gravitational force messenger particle computer. Then, the gravitational force messenger particle is transmitted, for example, to a down quark. The gravitational force messenger particle's embedded clock stores transmission time upon gravitational force messenger particle transmission. Upon gravitational force messenger particle reception, the embedded clock stores reception time. Also, the down quark's mass (m_2) is extracted from its associated Higgs force via the gravitational force messenger particle computer. The gravitational force messenger particle computer calculates the range (r) between the up quark and down quark as $r = (c)(\Delta t)$, where c is the velocity of light and Δt is the difference between reception and transmission times. Gravitational force between the two particles is calculated by the gravitational force messenger particle computer as $F = Gm_1m_2/r^2$, where G is the gravitational constant. Then, the gravitational force is provided to the receiving down quark.[24]

Matter particles and their associated Higgs forces were one and inseparable; that is, the three dimensional Higgs field was permanently attached to its associated matter particle. A model of a matter particle and its associated Higgs force is an underweight porcupine with overgrown spines as shown in figure 7. The underweight porcupine represents the matter particle (e.g., up quark) and the overgrown spines represent its associated three dimensional Higgs force field.

The Higgs mechanism was bidirectional, supporting condensations from and evaporations to the super force. That is, the matter particle and its associated Higgs force simultaneously condensed or froze out from a super force particle or simultaneously evaporated back to a super force particle. During matter creation, matter particles and their associated Higgs forces condensed from the super force as described in this chapter. As described in chapter 12, "Stellar Black Holes," and chapter 13, "Super Supermassive Quark Star (Matter)/Black Hole (Energy)," matter particles and their associated Higgs forces evaporated back to super force particles in our precursor universe's super supermassive quark star (matter) collapse to a super supermassive black hole (energy). A super supermassive quark star (matter) is a stellar black hole consisting of five fundamental matter particles (up quark, down quark, electron, zino, and photino) and their five associated Higgs forces. In contrast, a super supermassive black hole (energy) is a stellar black hole consisting entirely of super force energy particles.

The sum of five permanent Higgs force energies was dark energy as described in chapter 7, "Dark Energy."

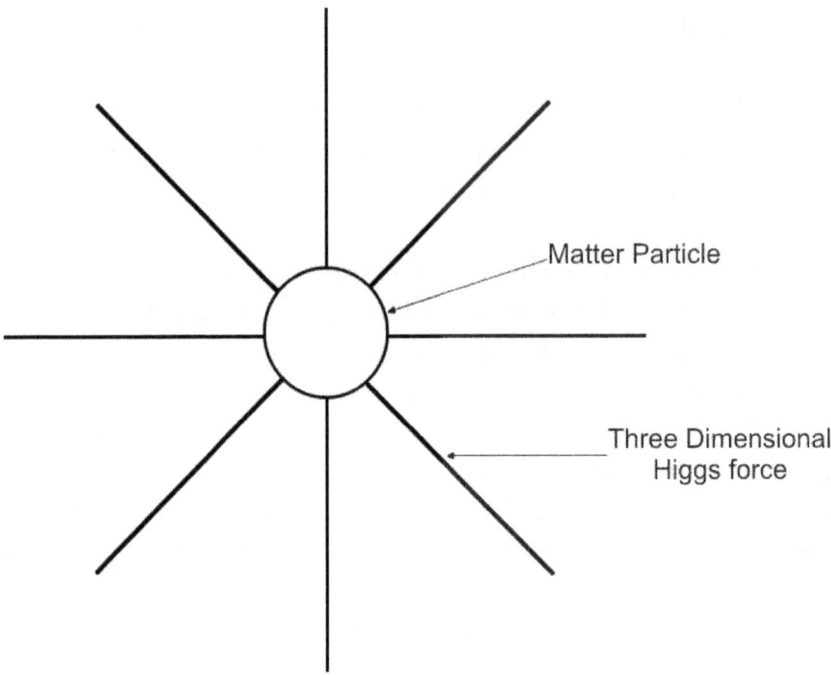

Figure 7. Matter particle (underweight porcupine) and its associated three dimensional Higgs force (overgrown spines).

Chapter 7

Dark Energy

The standard model of cosmology is the Lambda-Cold Dark Matter model. The sum of five Higgs force particles' energies associated with five permanent matter particles (down quark, up quark, electron, zino, and photino) is dark energy. The cosmological constant is proportional to dark energy density.

The standard model of cosmology is the Lambda-Cold Dark Matter (Λ-CDM) model,[25] where cosmology is the study of the origin, structure, and evolution of our universe. The Lambda-CDM model includes a big bang singularity and inflation. The big bang singularity is a point at the center of a single Planck cube, and inflation is an exponential increase in the size of our universe. At the start of our universe, all our universe's matter and force particles were in the form of near infinitely compressible super force particles in the singularity.[26] The Lambda-CDM model is consistent with the continuous expansion of our universe measured by astronomical observations. Dark energy is the unknown and proposed form of energy to explain our universe's expansion. Our universe decelerated during its first eight billion years and accelerated during the last six billion years. The accelerating rate is suggested by measurements of distances to galaxies using supernovas.[27] In the Lambda-CDM model, dark energy accounts for 72.6%, dark matter for 22.8%, and atomic matter for 4.6% of the total energy/mass of our universe. Mass and energy are interchangeable, like dollars and euros. The relationship between mass and energy is Einstein's equation $E = mc^2$, where E is energy, m is mass, and c is the velocity of light. For example, our universe's total energy/mass can be equivalently expressed as either 10^{54} kilograms of mass or 10^{90} electron volts of energy.

The sum of five Higgs force particles' energies associated with five permanent matter particles (down quark, up quark, electron, zino, and photino) was dark energy. The big bang singularity at the start of our universe, or time t = 0, was followed by an inflation, or exponential increase in the size of our universe. Inflation occurred immediately following the big bang singularity and caused the size of our universe to inflate from a sphere with a radius of .8 x 10^{-35} meters just touching the sides of a Planck cube, to a sphere with a radius of eight meters, or an exponential inflation factor of 8/.8 x 10^{-35} = 10^{36}. This inflation occurred in less than a millionth, of a billionth, of a billionth, of a billionth of a second, or 10^{-33} seconds. Particle creation occurred during and immediately following inflation. Five permanent matter particles and their five associated Higgs forces were created via their condensation from super force particles and exist today after 13.7 billion years. All the energy/mass of the super force singularity was converted via the Law of Conservation of Energy/Mass into atomic matter (up quark, down quark, electron), dark matter (zino, photino), and the Higgs forces or residual super force energies associated with the five permanent matter particles (up quark, down quark, electron, zino, photino). Thus the sum of the five residual super force energies, or the Higgs force particles' energies associated with the five permanent matter particles, was dark energy.

The cosmological constant was proportional to dark energy density. Following the end of matter creation, our universe consisted of dark energy (72.6%), dark matter (22.8%), and atomic matter (4.6 %), and those percentages remained constant for the next 13.7 billion years, or the age of our universe. Dark energy density was uniformly distributed in our universe and decreased with time as our universe expanded. The cosmological constant lambda (Λ) was proportional to dark energy density (ρ_Λ), or $\Lambda = k\rho_\Lambda$, where k is a constant. As our universe expanded, the cosmological constant also decreased along with dark energy density. Thus, the cosmological constant was not a constant but a variable which decreased as our universe expanded with time.

A model for our universe's dark energy was an expanding balloon filled with black smoke. Black smoke represented dark energy, and the expanding balloon represented our expanding universe. The amount of our universe's dark energy or black smoke was a constant during the age of our universe. Initially, the balloon, or our universe, was small in size, and the black smoke, or dark energy, was dense. As the balloon expanded, the black smoke density, or dark energy density, decreased. Since the cosmological constant was proportional to dark energy density, as our universe expanded, both dark energy density and the cosmological constant decreased.

Chapter 8

String Theory

There are five building block levels for matter and force particles. Strings are the fifth building block level for all five matter and six force particles. The energy/mass of a string is a function primarily of its diameter and secondarily its hill's and valley's amplitude displacement and frequency. The singularity at the start of our universe contained all of our universe's energy and consisted of superimposed super force strings.

There are five building block levels for matter and force particles, as shown in figure 8. The first four levels were described in previous chapters. In summary, there are five fundamental matter particles: three known atomic matter particles (up quark, down quark, and electron) and two proposed but undetected dark matter particles (zino and photino). There are six fundamental force or energy particles: four known (gravitational, electromagnetic, strong, and weak) and two proposed but undetected (super and Higgs). The five levels of matter and force particles are listed in the first column of figure 8. The second column shows the four building block levels of atomic matter: molecule; atom; proton/neutron and electron; and up quark, down quark, and electron. The electron is both a third and fourth level particle. In column three, dark matter is currently defined only at the fourth building block level or fundamental matter particles as the zino and photino. Conceptually, a periodic table of dark elements which defines nonatomic dark matter at upper levels exists but is currently undefined. The four known forces exist only at the fourth building block level, as shown in column four, and include gravitational, electromagnetic, strong, and weak. The two proposed but undetected force particles, super and Higgs, also exist only at the fourth level, as shown in columns five and six. There are five types of super forces which condense into five matter particles (up quark, down quark, electron, zino, and photino) and their five associated Higgs forces.

Strings are the fifth building block level for all five matter and six force particles, as shown in the fifth row of figure 8. All five matter particles (up quark, down quark, electron, zino and photino) and six force particles (gravitational, electromagnetic, strong, weak, five types of super, and five types of Higgs)[28] are strings. A string is visualized as a thin, sticky band wrapped around a Planck-cube-sized beach ball having surface hill's and valley's amplitude displacement and frequency. For example, the gravitational string shown in figure 9(a) has no hill's and valley's amplitude displacement and frequency and is a perfect circle with a Planck length diameter of 1.6×10^{-35} meters. In contrast, the up quark's string in Figure 9(b) is represented as a circle with hill's and valley's amplitude displacement and frequency.

The energy/mass of a string is a function primarily of its diameter and secondarily its hill's and valley's amplitude displacement and frequency. At the fifth string building blocks level, the only differences between the eleven matter and force particles are their hills' and valleys' amplitude displacement and frequency or their energy/mass. Figure 9(a) shows the gravitational force string, which is a perfect circular string wrapped around a Planck-cube-sized beach ball. A string with no hill's and valley's amplitude displacement and frequency represents zero energy. In contrast, the up quark string has hill's and valley's amplitude displacement and frequency as represented in figure 9(b), and its energy/mass is 2 million electron volts. The electromagnetic and strong force strings are perfect circles similar to the gravitational string and their energies are also zero. The down quark, electron, zino, photino, weak, Higgs, and super particles have energy/masses, so their strings are similar to the up quark string with different hills' and valleys' amplitude displacement and frequency.[29]

An artist's three-dimensional concept of an up quark in a Planck cube is shown in figure 10. The figure shows the up quark in the shape of a Planck-cube-sized beach ball with surface hill's and valley's amplitude displacement and frequency. The Planck-cube-sized beach ball with surface hill's and valley's amplitude displacement and frequency of figure 10 is the three-dimensional equivalent of the two-dimensional string of figure 9(b).[30] A cross section of the three-dimensional figure 10 at the YZ plane is the two-dimensional string of figure 9(b).

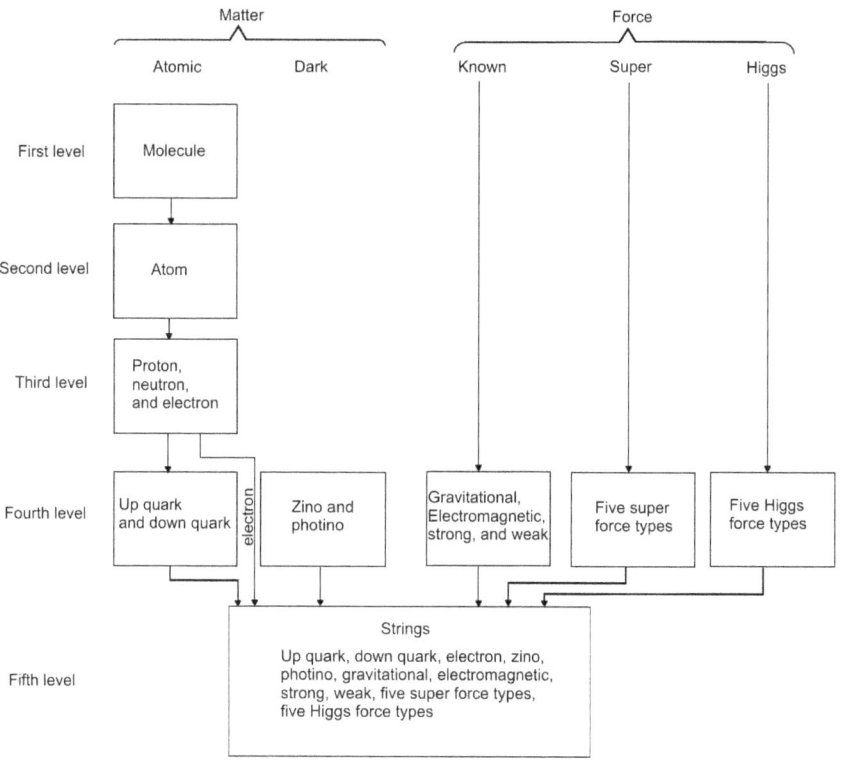

Figure 8. Five levels of matter and force particles.

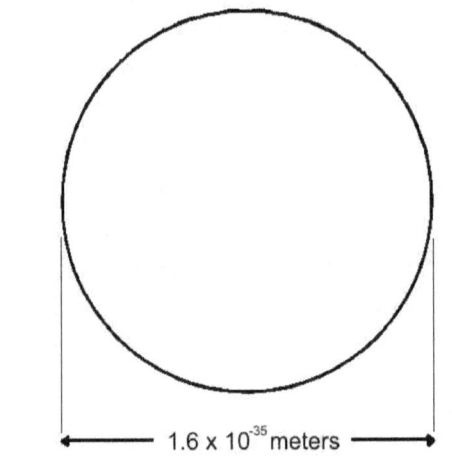

(a) Gravitational

Energy = 0

1.6 x 10^{-35} meters

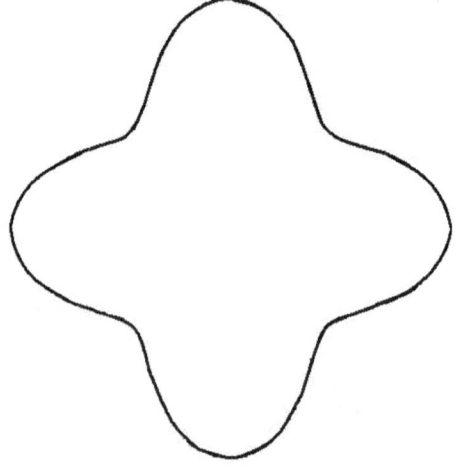

(b) Up quark

Energy/mass = 2 million electron volts

Figure 9. Particle strings.

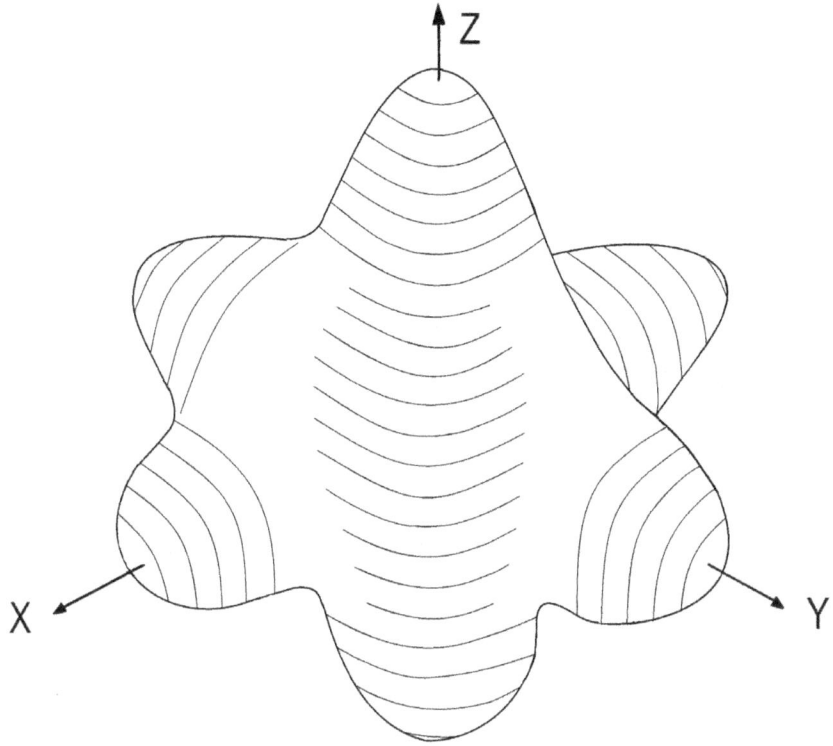

Figure 10. An up quark in three dimensions.

A Planck-cube-sized beach ball's potential energy/mass for each of the eleven fundamental matter and force particles can be modeled by three springs connected together at the Planck cube's center, as shown in figure 11. A beach ball just touching the Planck cube sides with no surface hill's and valley's amplitude displacement and frequency represents zero spring tension, or zero potential energy (e.g., the zero energy gravitational, electromagnetic, and strong forces). A range of amplitude displacements and frequencies about this zero energy level defines the energy/masses for nonzero energy/mass particles, such as the up quark, down quark, electron, zino, photino, weak, Higgs, and super.

The singularity at the start of our universe contained all of our universe's energy and consisted of superimposed super force strings. At the start of our universe, each of the five matter particles (up quark, down quark, electron, zino and photino), each of the four force particles (gravitational, electromagnetic, strong, and weak), and their nine associated Higgs particles were in the form of super force particles. Super force energy is intimately related to the energy/masses of its component five matter particles, four force particles, and their nine associated Higgs particles by the Law of Conservation of Energy/Mass.

The doughnut-shaped super force singularity at the center of a Planck cube is shown in figure 12. The super force singularity is a rotating, charged, doughnut-shaped, Kerr-Newman black hole. As described in chapter 12, "Stellar Black Holes," this singularity was created by our precursor universe's super supermassive quark star (matter) evaporation, deflation, and collapse to a super supermassive black hole (energy). A doughnut-shaped singularity's potential energy can also be modeled by three springs connected together at the Planck cube's center as shown in figure 11. Energy is a function inversely proportional to the singularity's diameter. The smaller the singularity's diameter, the greater is its potential energy in the compressed springs. At the start of our universe, our universe's energy of 10^{54} kilograms was in this doughnut-shaped singularity consisting of superimposed super force strings. If the three springs in figure 11 are compressed to the near zero diameter of the doughnut-shaped singularity of figure 12, the singularity's energy represents our universe's energy/mass of 10^{54} kilograms (mass), or 10^{90} electron volts (energy).

Near infinitely compressible super force particles were stacked one on top of another at the start of our universe.[31] At that time, everything in our universe was in the form of super force strings in the doughnut-shaped singularity. For example, you, I, New York city skyscrapers, the Great Wall of China, Mount Everest, the Rocky Mountains, all oceans, all seven billion world inhabitants, our solar system, all one hundred billion stars in our Milky Way galaxy, all one hundred billion galaxies in our universe, dark matter, and dark energy, were in the form of super force strings. At the start of our universe, all our universe's matter and force particles were superimposed super force strings in a doughnut-shaped singularity at the center of a single Planck cube.

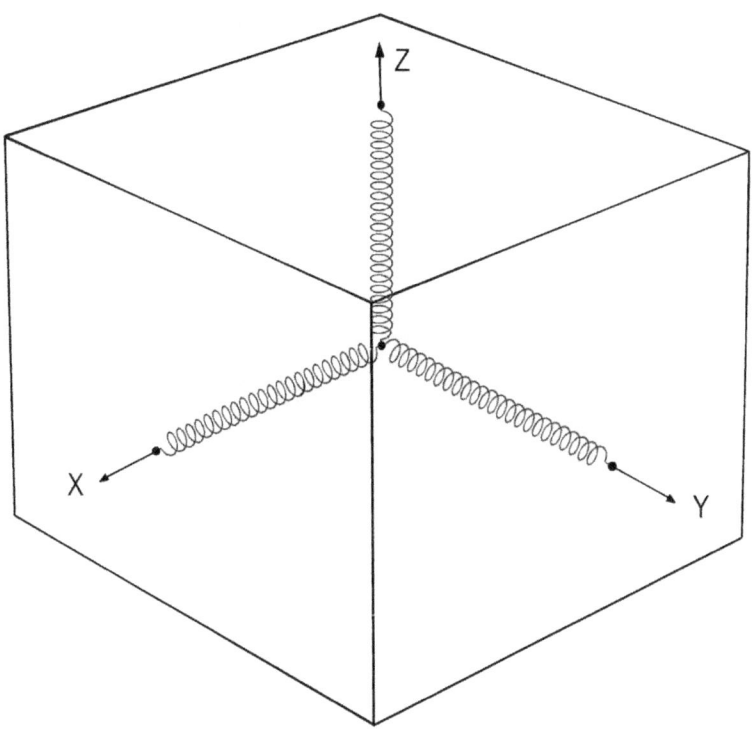

Figure 11. Three springs potential energy model.

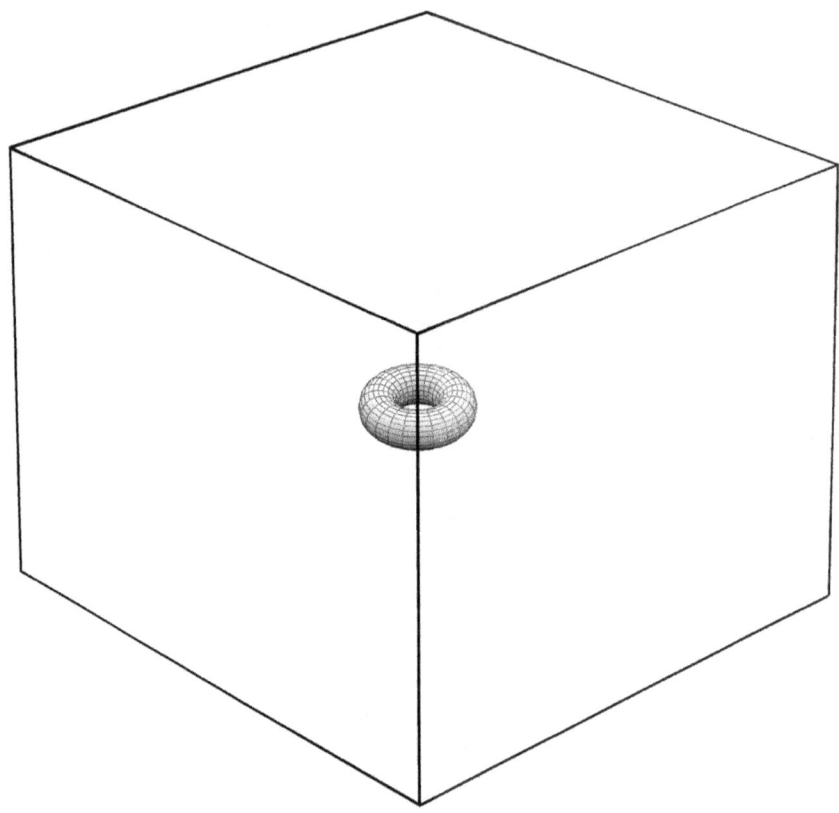

Figure 12. Big bang doughnut-shaped singularity in a Planck cube.

Chapter 9

Universe Expansions

Our universe had four time-sequential expansion phases:

1. Within our universe's first Planck cube;

2. Inflation;

3. Uniform distribution of matter;

4. Nonuniform distribution of matter.

Both the uniform and nonuniform distribution of matter expansions were driven by dark energy or Higgs force particles.

The first universe expansion occurred within our universe's first Planck cube. Our universe's size expanded from a doughnut-shaped singularity of near infinitely compressible super force particles at the center of a Planck cube to a spherical universe just touching the six sides of a Planck cube.[32] This phase occurred between the start of our universe at time t = 0 and the start of inflation at time t = 5 x 10^{-36} seconds, as shown in chapter 10, "Big Bang," figure 14, "Big bang time line." Since the Planck cube had edges equal to a Planck length of 1.6 x 10^{-35} meters, our universe's radius at the end of the first phase was one half a Planck length, or .8 x

10^{-35} meters. From that time on, our universe was an expanding spherical volume. This first expansion phase was similar to the unraveling of a smaller-than-Planck-cube knot of vibrating super force strings, or vibrating rubber bands.

The second universe expansion phase occurred during inflation and was similar to a water container freezing and bursting. More energy exists in liquid than frozen water. When water freezes, its temperature remains constant and latent heat is released.[33] At the start of inflation, six matter particles were created and expelled from the original Planck cube containing super force energy. Since matter particles must exist in their own Planck cubes, matter creation occurred only after our universe was larger than a Planck cube.[34] This process was described as the one to seven Planck cubes energy to matter expansion shown in figure 13. The center cube, hidden from view, contained our universe's super force energy (10^{54} kilograms), and a portion of that energy condensed or froze into the first six matter particles and their associated Higgs forces. In figure 13, the six matter cubes are explicitly shown and numbered as 2, 3, 4, 5, 6, and 7, whereas the center super force energy cube, numbered 1, is hidden from view. The first matter shell consisting of six matter particles was then pushed out, and a second matter Planck cube shell between the center Planck cube and the first matter shell was created. This process continued until enough shells with enough Planck cubes existed to accommodate all our universe's matter particles. By the end of inflation, our spherical universe had expanded from the size of a sphere just touching the sides of a Planck cube to a sphere with a radius of approximately 8 meters. The exponential inflation factor was 8 meters/.8 x 10^{-35} meters or 10^{36}, as shown in chapter 10, "Big Bang," figure 15, "Size of our universe versus time after the start of our universe."

The third universe expansion phase, or the uniform distribution of matter, occurred from the end of inflation to 30,000 years after the start of our universe. This is shown in chapter 11, "Our Universe," figure 17, "Our universe's time line." Uniform distribution meant matter was uniformly distributed in our universe. For example, samples anywhere in our universe contained the same number of matter particles.

At 30,000 years after the start of our universe, dark matter began to clump together, and this was the beginning of our universe's nonuniform distribution of matter. These dark matter clumps formed the framework of our universe's galaxies and began the galaxy evolution process. Over 100 billion clumps of dark matter occurred in our early universe, and these were the framework nodes or three-dimensional locations of our universe's 100 billion galaxies. As described in chapter 3, "Two Dark Matter Particles," the period between 30,000 and 380,000 years was a transition period

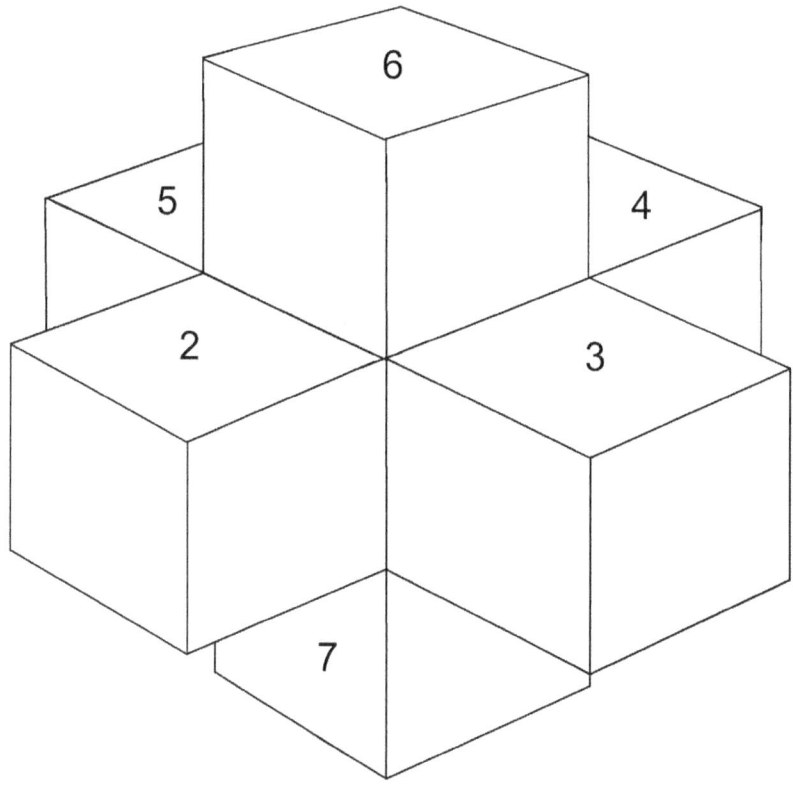

Figure 13. One to seven Planck cubes energy to matter expansion.

between uniform distribution of matter and nonuniform distribution of matter. During the transition period, atomic matter (up quarks, down quarks, and electrons) was uniformly distributed, whereas dark matter (zinos and photinos) was not.

The fourth universe expansion phase, or the nonuniform distribution of matter expansion, occurred from 30,000 years after the start of our universe to the present time. Nonuniform distribution of matter meant matter was not uniformly distributed; for example, galaxies were lumps of atomic and dark matter surrounded by space. Following creation of galaxies at 200 million years, as shown in figure 17, "Our universe's time line," our universe's nonuniform distribution of matter expansion could be represented by a marbles/rising dough/balloon model. The marbles were mixed in rising dough in an expanding balloon. The rigid marbles represented galaxies which did not expand in size and were nonuniformly distributed in the balloon. The rising dough represented intergalactic space, or space between galaxies. The dough, or intergalactic space, rose or expanded with time. The spherical balloon represented our universe, which expanded with time along with intergalactic space.

Both the uniform and nonuniform distributions of matter expansions were driven by dark energy or Higgs force particles. Similar to the expansion of carbon dioxide gas in a room after a bottle of Coke is opened, dark energy or Higgs force particles drove both the uniform and nonuniform distribution of matter expansions.[35] Following 13.7 billion years of expansion, our spherical universe presently has a radius of approximately 46.5 billion light-years, or 440 billion trillion kilometers.

Chapter 10

Big Bang

Matter creation was the condensation of matter particles and their associated Higgs forces from the super force or mother particle. There were five types of super force particles. Matter creation was time synchronous with both the inflationary period start time and the one to seven Planck cubes energy to matter expansion. Our universe expanded from a singularity at the start of our universe to a hot particle soup by the end of inflation. Our universe had three atom production processes.

Matter creation was the condensation of matter particles and their associated Higgs forces from the super force or mother particle. All our universe's superimposed super force strings were in a doughnut-shaped singularity at the start of our universe or big bang at time t = 0. As shown in the "Big bang time line" of figure 14, the start and end of matter creation occurred during the time interval 5×10^{-36} to 100 seconds. The start of matter creation coincided with the start of inflation. Inflation occurred during the time interval 5×10^{-36} to 10^{-33} seconds. Similar to water having three phases, steam, water, and ice, the super force had nine additional phases—up quark, down quark, electron, zino, photino, gravitational, electromagnetic, strong, and weak—and their nine associated Higgs particles. Just as steam condensed to water at 212^0 Fahrenheit, and water condensed or froze to ice at 32^0 Fahrenheit, the super force condensed at nine different and extremely high temperatures between 10^{27} and 10^{10}

degrees Kelvin to the five matter particles, four force particles, and their nine associated Higgs particles.[36]

At the start of our universe, our universe was a doughnut-shaped super force singularity at the center of a single Planck cube as shown in figure 12, "Big bang doughnut-shaped singularity in a Planck cube." The super force doughnut-shaped singularity had near infinite energy density. That is, density equals energy/volume, where the energy equaled 10^{54} kilograms divided by the near zero volume of the doughnut-shaped singularity. During matter creation, the super force singularity was condensed into a near infinite number of fundamental matter particles (up quark, down quark, electron, zino, and photino).[37] Each fundamental matter particle was in a Planck cube as shown in figure 9(b), "Up quark particle string," and figure 10, "An up quark in three dimensions." Creation of matter and force particles from the doughnut-shaped super force singularity can be modeled as a wand made of soap and water. The wand represented the doughnut-shaped super force singularity which created permanent bubbles of matter and force particles. The wand dissolved itself during condensation of all super force particles (soap and water) into matter and force particles (permanent bubbles).

There were five types of super force particles. A super force particle and each of the condensed five matter particles with their associated Higgs forces had equal energy/masses. By the Law of Conservation of Energy/Mass, the energy of a super force particle equaled the energy/mass of its products, namely, the condensed matter particle and its associated Higgs force. Two examples of super force types were the super force that created the up quark and its associated Higgs force and the super force that created the down quark and its associated Higgs force. Since the five matter particles and their associated Higgs forces had different energy/masses, their originating five super force particles had different energies.

Matter creation was time synchronous with both the inflationary period start time and the one to seven Planck cubes energy to matter expansion as described in chapter 9, "Universe Expansions," and figure 13, "One to seven Planck cubes energy to matter expansion." Matter particles did not exist before our universe was the size of a single Planck cube. As shown in figure 15, "Size of our universe versus time after the start of our universe,"[38] by the horizontal dash and point line, the start of inflation at 5 x 10^{-36} seconds occurred when the radius of our universe was .8 x 10^{-36} meters or the size of a Planck cube. The one to seven Planck cubes expansion consisted of six contiguous Planck cubes containing six matter particles, attached to the six faces of our universe's original Planck cube containing super force particles. Our universe expanded from the size of a sphere just touching the sides of a Planck cube at the start of inflation to a spherical volume with a radius of 8 meters at the end of inflation.

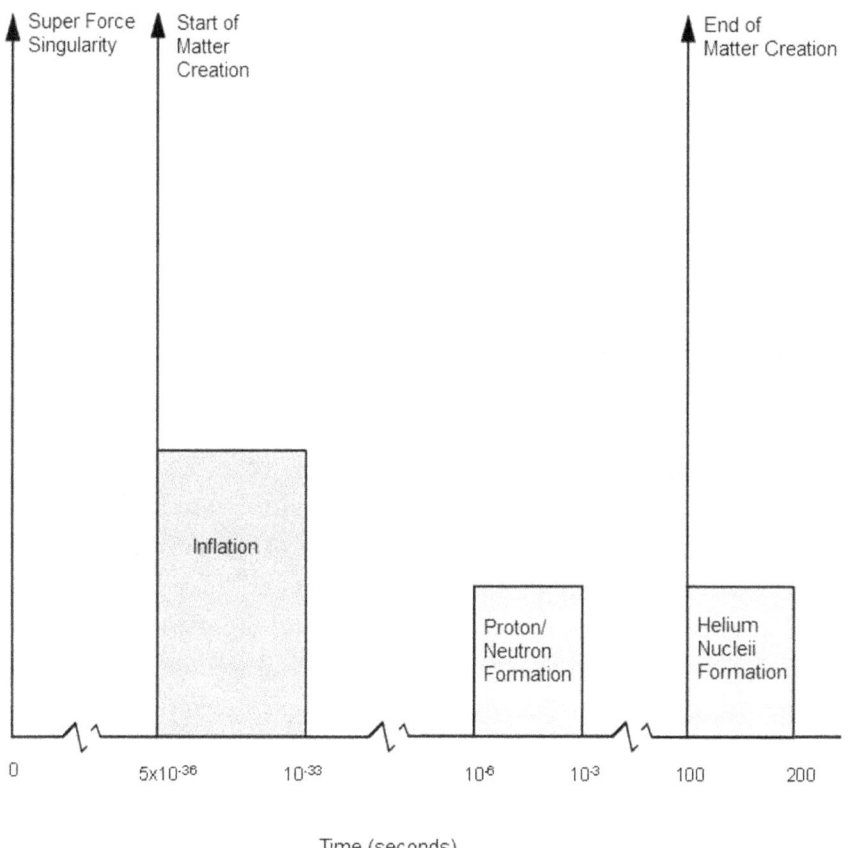

Figure 14. Big bang time line.

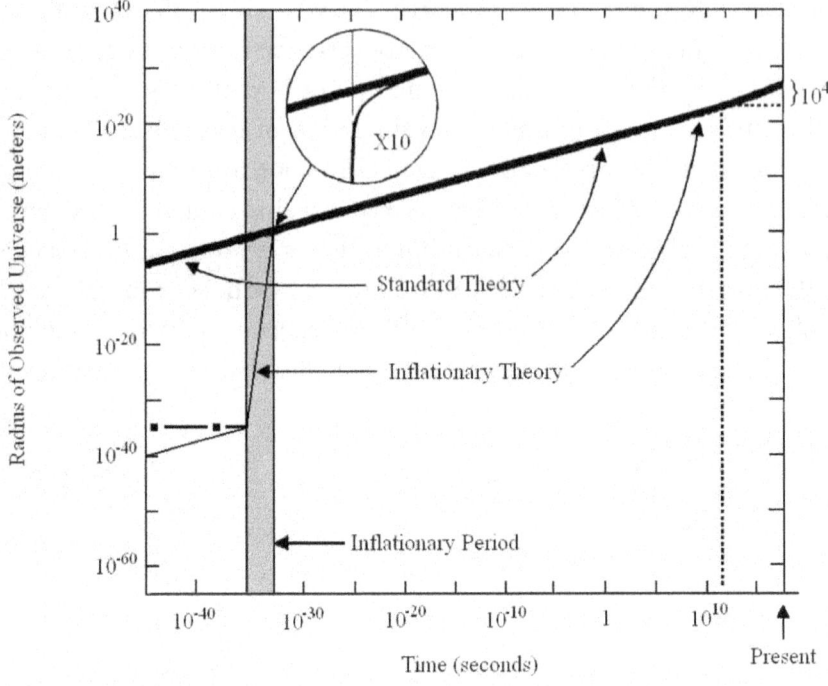

Figure 15. Size of our universe versus time after the start of our universe.

Our universe expanded from a singularity at the start of our universe to a hot particle soup[39] by the end of inflation, as shown in figure 16. The hot particle soup consisted of a subset of the up quark, down quark, electron, zino, and photino matter particles. At the start of our universe, our universe consisted of a doughnut-shaped singularity at the center of a single Planck cube as shown in figure 16(a). By the end of inflation at t = 10^{-33} seconds, our universe had expanded to a spherical volume with a radius of 8 meters and contained a hot particle soup, as shown in figure 16(b). Figure 16(b) is shown in two, not three, dimensions and not to scale, since the Planck cubes are much smaller than our universe with an 8-meter radius. A matter particle in its Planck cube is shown as an m and a Higgs force as an h in figure 16(b). The three-dimensional Higgs forces associated with each matter particle occupied all Planck cubes not containing a matter particle. For simplicity, the four other force particles (gravitational, electromagnetic, strong, and weak) existed but were not explicitly shown in the hot particle soup of figure 16(b).

Our universe had three atom production processes. During the proton and neutron formation period[40] between 10^{-6} and 10^{-3} seconds, shown in figure 14, protons and neutrons formed when two up quarks and one down quark combined to form protons and one up quark and two down quarks combined to form neutrons. By the end of matter creation at t = 100 seconds, all super force particles had condensed to matter particles and their associated Higgs forces. Following the end of matter creation, the first of our universe's three atom production processes occurred. During the big bang helium nuclei formation between 100 and 200 seconds of figure 14, two protons and two neutrons combined to form nuclei of helium atoms.[41] The second and third atom production processes, stellar core and supernova explosion, are described in chapter 11, "Our Universe."

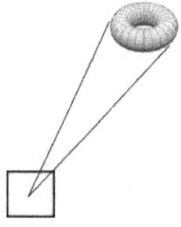

(a) Singularity

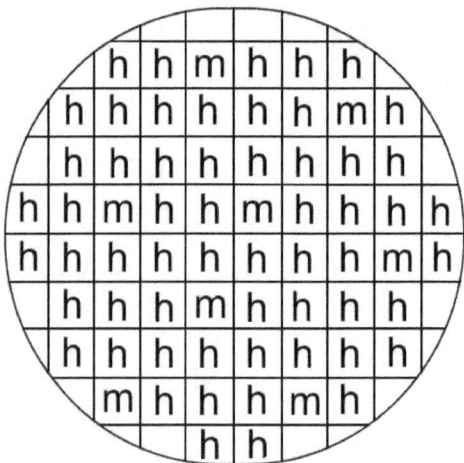

Radius = 8 meters

(b) Hot particle soup

Figure 16. Our universe's expansion from a singularity to a hot particle soup.

Chapter 11

Our Universe

Our universe started with the big bang, and matter creation occurred between 5 x 10^{-36} and 100 seconds as described previously in chapter 10, "Big Bang." Dark matter began clumping at our universe's three-dimensional framework nodes at 30,000 years.[42] At 380,000 years, atomic matter consisting of neutral hydrogen and helium atoms began to clump around the dark matter nodes. First-generation stars[43] were created by hydrogen and helium molecular clouds collapsing at these dark matter framework nodes starting at 200 million years. First-generation stars gravitationally collapsed either to neutron stars followed by supernova explosions or quark stars (matter) followed by quark-nova explosions. The supernova and quark-nova explosions created primeval galaxies[44] associated with neutron stars and quark stars (matter), respectively. By accretion of stars/matter and merger with other galaxies, these early neutron and quark stars (matter) and their associated primeval galaxies evolved into our present universe's approximately 100 billion supermassive quark stars (matter) and their associated galaxies. Elements heavier than hydrogen and helium were created by the second and third atom production processes, stellar core and supernova explosion. In the stellar core atom production process, elements from carbon to iron were created. In the supernova atom production process, all remaining elements heavier than iron were created. Stellar core and supernova explosion atom production processes created all our universe's atoms except hydrogen and helium.

Water- and carbon-based molecules were required for life. Our solar system was created at 9.1 billion years and the present time is 13.7 billion years after the start of our universe.

Dark matter began clumping at our universe's three-dimensional framework nodes at 30,000 years, as shown in our universe's time line of figure 17. Dark matter (i.e., zino and photino) clumping created the three-dimensional framework nodes of our universe's galaxies. Over 100 billion dark matter clumps were created. Dark matter clumped between 30,000 and 380,000 years after the start of our universe, whereas electrically charged atomic matter such as hydrogen nuclei, helium nuclei, and electrons, did not. Hydrogen nuclei consisted of positively charged protons, helium nuclei consisted of two positively charged protons and two neutral neutrons, and electrons were negatively charged. Atomic matter did not clump together because the high kinetic energy or temperature of the hydrogen nuclei, helium nuclei, and electrons outweighed their electromagnetic attraction forces.

At 380,000 years, atomic matter consisting of neutral hydrogen and helium atoms began to clump around the dark matter nodes. By 380,000 years, our universe had cooled sufficiently to allow positively charged hydrogen and helium nuclei to combine with negatively charged electrons to form neutral hydrogen and helium atoms.[45]

First-generation stars were created by hydrogen and helium molecular clouds collapsing at these dark matter framework nodes starting at 200 million years. First-generation stars contained up to 100 times more hydrogen and helium gas than our sun, had short lives, exploded as supernovas or quark-novas, and created over 100 billion neutron and quark stars (matter) and their associated primeval galaxies. If the first-generation star was between 8 and 20 times the mass of our sun or solar masses, it collapsed to a dense neutron star. If the first-generation star was between 20 and 100 solar masses, it first collapsed to a dense neutron star followed by a delayed collapse to a denser quark star (matter). The remnants of supernovas or quark-novas were the associated primeval galaxies.

First-generation stars gravitationally collapsed either to neutron stars followed by supernova explosions or quark stars (matter) followed by quark-nova explosions. The supernova and quark-nova explosions created primeval galaxies associated with neutron stars and quark stars (matter), respectively. A supernova was the gravitational collapse of a massive star to a dense neutron star, the subsequent explosion which released enormous matter and radiation energy, and the formation of its associated primeval galaxy. A neutron star was a dense star consisting primarily of neutrons. A quark-nova was the gravitational collapse of a neutron star to an even denser quark star (matter), the subsequent explosion which released stupendous matter and radiation energy, and the formation of its associated primeval galaxy. All of the neutron's strong force energy confining the up quarks and down quarks in the neutron was released during the quark-nova explosion.[46] A quark star (matter) was denser than a neutron star and consisted of five fundamental matter particles (up quark, down quark, electron, zino, and photino).

By accretion of stars/matter and merger with other galaxies, these early neutron and quark stars (matter) and their associated primeval galaxies evolved into our present universe's approximately 100 billion supermassive quark stars (matter) and their associated galaxies. First-generation star collapses created over 100 billion neutron and quark stars (matter) and their associated primeval galaxies. These neutron stars and quark stars (matter) and their associated primeval galaxies increased in size via accretion of stars, matter between stars,[47] matter between galaxies,[48] and by merger with other galaxies. Currently, 13.7 billion years after the start of our universe, approximately 100 billion galaxies exist with 100 billion supermassive quark stars (matter) at their centers. These supermassive quark stars (matter) range in size from 1 million to 10 billion solar masses.

Elements in the periodic table of elements heavier than hydrogen and helium[49] were created by the second and third atom production processes, stellar core and supernova explosion.[50] Both processes involved the weak force or the alchemist of the universe to transform an atom into a different type of atom.

Master Big Bangs through Black Holes in Four Hours

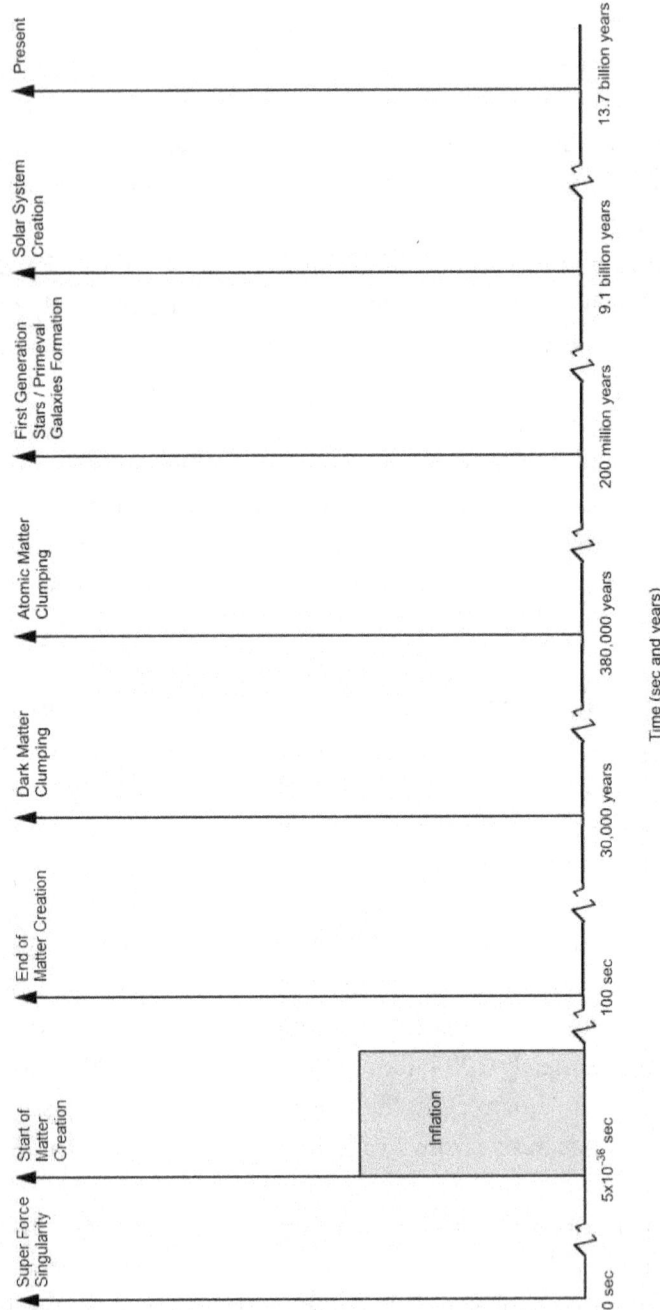

Figure 17. Our universe's time line.

In the stellar core atom production process, elements from carbon to iron were created. Hydrogen in the stellar core first fused into helium and released energy. The released energy provided sufficient pressure to prevent the stellar core from collapsing. Following hydrogen exhaustion, helium in the stellar core fused into carbon and released energy. The released energy provided sufficient pressure to prevent the stellar core from collapsing. This cycle repeated itself, and each time the core began to collapse, its temperature and pressure increased and more massive nuclei fused and released energy. At the end of these fusion processes, the massive star core resembled a layered onion shown in figure 18, "Stellar core atom production." The most massive iron nuclei were in the star's core, and each higher layer contained less massive nuclei, with the least massive hydrogen nuclei in the top layer.

In the supernova atom production process, all remaining elements heavier than iron were created. Since iron did not fuse, there was no released energy to provide sufficient pressure to prevent the stellar iron core of figure 18 from collapsing. This initiated the third atom production process, called supernova explosion. In this process, the iron core gravitationally collapsed into a dense neutron star. The star core's upper layers from silicon to hydrogen literally had the floor dropped from underneath them, so these upper layer atoms began to accelerate toward the newly created dense neutron star. The accelerating atoms were stopped by the dense neutron star, bounced off, and produced an outwardly propagating shock wave. In this supernova explosion, energetic neutrons were produced which bombarded nuclei. These nuclei built up atomic mass one neutron at a time to produce heavier unstable nuclei. These unstable nuclei transformed via the weak force to stable nuclei. That is, new chemical elements or atoms were formed with atomic numbers heavier than nickel.[51]

Stellar core and supernova explosion atom production processes created all our universe's atoms except hydrogen and helium. Supernova and quark-nova remnants were molecular clouds[52] and contained hydrogen atoms, helium atoms, and heavier elements used for the creation of new stars and planets. Thus, molecular clouds were both the graveyards of stars (e.g., collapse of a first-generation star) and the birthplaces of new stars.

Water- and carbon-based molecules were required for life. Water molecules consisted of two hydrogen atoms and one oxygen atom. Carbon-based molecules consisted of carbon, hydrogen, oxygen, and nitrogen atoms.[53] Water on earth was delivered by comets and asteroids, which contained water in the form of ice. Comets, which consisted primarily of ice, and asteroids, which consisted primarily of rocks, metals, and ice, bombarded the earth during its formation following our solar system's creation at 9.1 billion years. Water presence in galaxies has been detected via distinctive water fingerprint frequencies.[54]

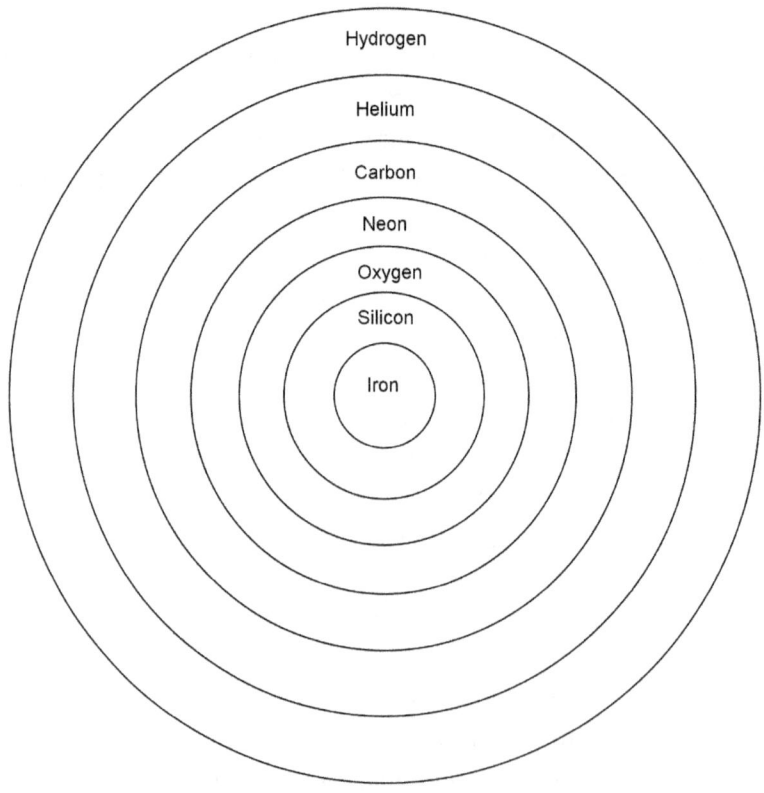

Figure 18. Stellar core atom production.

Carbon molecules were also formed in molecular clouds following supernova explosions. Recent astronomical observations revealed the presence of carbon molecules in all galaxies and that they were detected in their planets, comets, and asteroids. Comets and asteroids containing carbon molecules bombarded the earth during its formation following our solar system's creation. From these carbon molecules, primitive life in the form of simple cells was created.

Our solar system was created at 9.1 billion years, and the present time is 13.7 billion years after the start of our universe. At our solar system's creation 4.6 billion years ago, a molecular cloud containing hydrogen and helium gas, dust, and charged particles[55] began ordering itself. A molecular cloud gravitationally collapsed with most of its mass going into our sun and the remaining mass into a disk containing planets, moons, asteroids, and comets. Simple life cells were created approximately 3.5 billion years ago. These simple life cells evolved into multicell life, which in turn evolved into fish, amphibians, and land mammals. The first animals appeared approximately 800 million years ago and consist today of an estimated seven million species.

The first modern humans appeared approximately 200 thousand years ago in East Africa. There were many waves of human migration. In one wave, modern humans migrated north from East Africa via the Arabian Peninsula to the Near East 125,000 years ago. One group migrated west from the Near East and reached Europe 40,000 years ago. A second group from the Near East migrated east along the Asian coast reaching South East Asia 50,000 years ago. A subset of the second group reached Australia 40,000 years ago. A second subset of the second group from the Near East migrated northeast to Central Asia. The Americas were populated by a subset of this group between 30,000 and 15,000 years ago.

In a second wave approximately 7,500 years ago, an ice age ended and the level of the oceans and Mediterranean Sea rose. Originally the Black Sea was a freshwater lake, but the waters of the Mediterranean Sea rose above the dammed Bosporus Strait and converted the Black Sea from fresh to salt water. In Crimea, an area the size of Florida flooded and caused three human migrations. One group went south, crossed the Black Sea in arks described in the biblical story of Noah's Ark,[56] and settled in Mesopotamia, now known as Iraq. A second group went east and populated India, China, and Southeast Asia. A third group went west and settled Greece, Europe, and eventually Great Britain.

Our universe has evolved since the big bang. Thus, 13.7 billion years after the start of our universe, we and our world exist.

Chapter 12

Stellar Black Holes

Quark stars (matter) and black holes (energy) are stellar black holes which are so dense not even light can escape from them. Six types of stellar black holes are quark stars (matter), supermassive quark stars (matter),[57] super supermassive quark stars (matter), their associated super supermassive black holes (energy), super super supermassive quark stars (matter), and their associated super super supermassive black holes (energy).[58] In our precursor universe and at the 10^{24} solar masses threshold, a super supermassive quark star (matter) evaporated, deflated, and collapsed to its associated super supermassive black hole (energy) and created the big bang of our universe. Our precursor universe's super supermassive quark star (matter)/black hole (energy) was a gigantic compactor.

Quark stars (matter) and black holes (energy) are stellar black holes which are so dense not even light can escape from them. The current definition of a stellar black hole has inconsistencies. For example, on one hand, a stellar black hole is defined as a singularity which has near zero surface area or volume. On the other hand, a stellar black hole is defined with maximum disorder directly proportional to surface area or volume. Because of these inconsistencies, stellar black hole theory was amplified to define a stellar black hole as either a quark star (matter) or black hole (energy), both of which are "black." Their differences are that a quark star (matter) has mass, volume, near zero temperature, permanence, and maximum disorder. In contrast, a black hole (energy) has no mass, a Planck cube singularity with minimal volume, near infinite temperature, transientness, and minimal disorder, as described in detail in chapter 13, "Super Supermassive Quark Star (Matter)/Black Hole (Energy)."

The first type of stellar black hole is the quark star (matter) which contains between several and one million solar masses. For example, quark stars (matter) having several solar masses were initially created by first-generation star collapses beginning approximately 200 million years after the start of our universe. Their sizes were augmented by accretion of stars/matter and merger with neutron star or quark star (matter) galaxies during the next 13.5 billion years.

The second type of stellar black hole is the supermassive quark star (matter), which contains between a million and 10 billion solar masses. For example, quark stars (matter) having several solar masses were initially created by first-generation star collapses beginning approximately 200 million years after the start of our universe. Over the next 13.5 billion years, by accretion of stars/matter and merger with galaxies, approximately 100 billion supermassive quark stars (matter) and their 100 billion galaxies formed in our universe.[59] That is, over the last 13.5 billion years, between a million and 10 billion solar masses fell into the original neutron or quark stars (matter). At the center of each of the 100 billion galaxies in our universe is a supermassive quark star (matter). The supermassive quark star (matter) at the center of our Milky Way galaxy is called Sagittarius A. The supermassive quark star (matter) is a three-dimensional sphere which appears as a two-dimensional black hole from any viewing angle. This is similar to our three-dimensional sun, which appears as a two-dimensional disk from any viewing angle.

A supermassive quark star (matter) is to stars in our Milky Way galaxy as our sun is to planets in our solar system. Just as the sun's gravity holds the planets rotating about it, the gravity of the Milky Way's supermassive quark star (matter) holds the stars rotating about it. Currently, the energy/mass of the Milky Way's supermassive quark star's (matter) is estimated at

Stellar Black Holes

4 million solar masses. Since a supermassive quark star (matter) is a stellar black hole so dense not even light can escape from it, simulated images of spiral galaxies[60] similar to our Milky Way galaxy show stars rotating around the black circular disk of its supermassive quark star (matter). The first two types of stellar black holes, namely, quark star (matter) and supermassive quark star (matter) are not large enough to collapse and form an associated black hole (energy).

The third and fourth types of stellar black holes are the super supermassive quark star (matter) and its associated super supermassive black hole (energy). A super supermassive quark star (matter) contains between 10 billion solar masses and our universe's energy/mass, or 10^{24} solar masses. The similarities between a super supermassive quark star (matter) and its associated super supermassive black hole (energy) are both had the same energy/mass and both existed in our precursor universe. In our precursor universe, a super supermassive quark star (matter) increased in size via accretion of stars/matter and merger with galaxies.

In our precursor universe and at the 10^{24} solar masses threshold, a super supermassive quark star (matter) evaporated, deflated, and collapsed to its associated super supermassive black hole (energy) and created the big bang of our universe. At the 10^{24} solar mass threshold, quark pressure was insufficient to stop further collapse, and the super supermassive quark star (matter) collapsed to its associated super supermassive black hole (energy). The super supermassive quark star (matter) instantaneously evaporated, deflated, and collapsed to the doughnut-shaped singularity of the super supermassive black hole (energy) shown in figure 19. As shown in figure 19(a), the extremely large super supermassive quark star (matter) existed at approximately -1 second, or one second before the start of our universe. Its radius was estimated as much less than 10^{26} meters, or $<< 10^{26}$, and the deflation time as approximately 2×10^{-33} seconds.[61] Figure 19(a) shows matter particles (up quark, down quark, electron, zino, and photino) and Higgs forces within Planck cubes. A matter particle in its Planck cube is shown as an m and a Higgs force as an h in figure 19(a). Figure 19(a) is shown in two, not three, dimensions and is not to scale

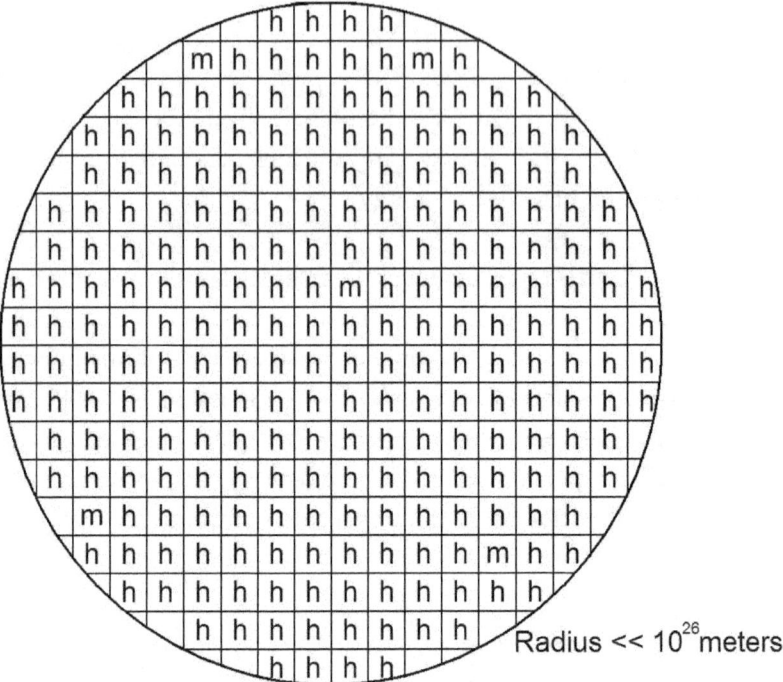

(a) Super supermassive quark star (matter) at t=-1 second

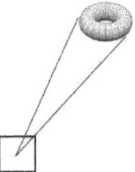

(b) Super supermassive black hole (energy) at t=0

Figure 19. Super supermassive quark star (matter) collapse to a super supermassive black hole (energy).

since Planck cubes are much smaller than the super supermassive quark star (matter) with a radius << 10^{26} meters.

At the center of the super supermassive quark star (matter), a single electron and its associated Higgs force were subjected to the highest pressure or temperature. This electron and its associated Higgs force evaporated to the super force, incrementally raising the temperature of the super supermassive quark star (matter) center. This began a chain reaction which instantaneously evaporated, deflated, and collapsed the super supermassive quark star (matter) to a super supermassive black hole (energy). The latter is shown in figure 19(b) as a doughnut-shaped super force singularity at the center of a single Planck cube. The deflation time was estimated to be approximately twice our universe's big bang inflation time, or approximately 2×10^{-33} seconds. Our precursor universe's super supermassive quark star (matter)/black hole (energy) created our universe's "big bang" (white hole) via Conservation of Energy/Mass. A white hole was the reverse of a black hole. A black hole swallowed matter and energy, whereas a white hole expelled it. Super supermassive quark stars (matter)/black holes (energy) were to universes as supermassive quark stars (matter) were to galaxies.

The fifth and sixth types of stellar black holes are the super super supermassive quark star (matter) and its associated super super supermassive black hole (energy). A super super supermassive quark star (matter)/black hole (energy) contains much greater than 10^{24} solar masses. A super super supermassive quark star (matter) evaporated, deflated, and collapsed to its associated super super supermassive black hole (energy) and created our precursor universe. Super super supermassive quark stars (matter)/black holes (energy) were to precursor universes as super supermassive quark stars (matter)/black holes (energy) were to universes.

Our precursor universe's super supermassive quark star (matter)/black hole (energy) was a gigantic compactor. Imagine a gigantic compactor similar to a junkyard compactor that compresses car chassis into steel cubes. Our imagined gigantic compactor can compress any large object into a near infinitely small volume. For example, imagine everything in New York City, including the Empire State Building, Radio City Music Hall, the Brooklyn Bridge, and the Verrazano Bridge, is placed into the gigantic compactor and compressed. Then, everything in the United States, including the Appalachian Mountains, Gateway Arch, the Great Lakes, the Rocky Mountains, and Los Angeles, is placed into the gigantic compactor and compressed. Everything in the world, including the Great Wall of China, Mount Everest, Mount Kilimanjaro, the Atlantic, Pacific, and Indian Oceans, and all seven billion people, is placed into the gigantic compactor and compressed. Everything in our solar system, including the sun, all planets, all moons, asteroids,[62] and comets,[63] is placed into the gigantic compactor and compressed. Everything in our Milky Way galaxy, including all one hundred billion stars, each with their planets, moons, asteroids, and comets, is placed into the gigantic compactor and compressed. All one hundred billion galaxies in our universe, each with one hundred billion stars and each star with its planets, moons, asteroids, and comets, is placed into the gigantic compactor and compressed. All our universe's dark matter and dark energy is also placed into the gigantic compactor and compressed. Our imagined gigantic compactor compressed all of our universe's matter and energy into a doughnut-shaped super force singularity at the center of a single Planck cube. This happened just prior to the start of our universe, except our precursor universe's subset volume objects were compressed instead of our universe's objects. Our precursor universe had the same fundamental matter particles (up quark, down quark, electron, zino, and photino) and fundamental force particles (gravitational, electromagnetic, strong, weak, super, and Higgs) as our universe. However, the molecular, atomic, and nuclear objects of our precursor universe's subset volume were not necessarily the same as our universe's objects. For example, the specific Empire State Building, which consists of a unique structure of molecules, atoms, up quarks, down quarks, and electrons in our universe, may not have existed in our precursor universe's subset volume.

Our described gigantic compactor is our precursor universe's super supermassive quark star (matter)/black hole (energy). Black holes compress matter as described above, and their operation is well documented.

However, our gigantic compactor description is a revolutionary idea, because it describes a super supermassive quark star (matter) which collapsed into a super supermassive black hole (energy), which created the big bang of our universe. Matter creation in our universe's big bang occurred during and after inflation. Super force particles condensed into five matter particles, four force particles, and their nine associated Higgs particles. The reverse of this condensation occurred in our precursor universe's subset volume during the evaporation, deflation, and collapse of the super supermassive quark star (matter) to the super supermassive black hole (energy). The five matter particles, four force particles, and their nine associated Higgs particles in the super supermassive quark star (matter) instantaneously evaporated, deflated, and collapsed into super force particles of the super supermassive black hole (energy) as shown in figure 12. The super supermassive quark star (matter)/black hole (energy) had the same energy/mass of our universe, or 10^{54} kilograms, or 10^{24} solar masses.[64]

Chapter 13

Super Supermassive Quark Star (Matter)/ Black Hole (Energy)

Our precursor universe's super supermassive quark star (matter) and its associated super supermassive black hole (energy) had identical energy/masses. Their differences involved mass, volume, temperature, longevity, and disorder. Three laws of physics determined the existence or nonexistence of precursor universes and super supermassive quark stars (matter)/black holes (energy). The super supermassive quark star (matter) to black hole (energy) collapse in a subset volume of our precursor universe reset disorder from maximum to minimum and "resurrected" life via creation of the super force or mother particles.

Our precursor universe's super supermassive quark star (matter) and its associated super supermassive black hole (energy) had identical energy/masses. Our precursor universe's super supermassive quark stars (matter) were stellar black holes having energy/masses between 10 billion (10^{10}) and our universe's energy/mass of 10^{24} solar masses. At the 10^{24} solar masses threshold, our precursor universe's super supermassive quark star (matter) collapsed to a super supermassive black hole (energy) and created our universe. Because of the Law of Conservation of Energy/

Mass, both the super supermassive quark star (matter) and its associated super supermassive black hole (energy) had identical energy/masses of 10^{24} solar masses.

Differences between a super supermassive quark star (matter) and its associated super supermassive black hole (energy) are that the former has mass, volume, near zero temperature, permanence, and maximum disorder, whereas the latter has no mass, a Planck cube singularity with minimum volume, near infinite temperature, transientness, and minimum disorder.

The super supermassive quark star (matter) had mass because it consisted of a near infinite number of fundamental matter particles (up quarks, down quarks, electrons, zinos, and photinos). It had a large volume whose estimated radius was much less than 10^{26} meters ($<< 10^{26}$), shown in figure 19(a). Its temperature was near zero and it had permanence.[65] Its disorder was a maximum.[66] In contrast, the super supermassive black hole (energy) had no mass because it consisted only of super force energy particles. It volume was near zero since it was a doughnut-shaped singularity at the center of a single Planck cube. Its temperature was near infinite and it had transientness. The doughnut-shaped singularity of super force particles was transient because by the end of matter creation at 100 seconds, the singularity had condensed to atomic matter, dark matter, and dark energy. At time = 0, its disorder was a minimum because it consisted of superimposed super force particles in the same place, a singularity.

Three laws of physics determine the existence or nonexistence of precursor universes and super supermassive quark stars (matter)/black holes (energy). The three laws of physics are Conservation of Energy/Mass, Einstein's Theory of General Relativity, and the Second Law of Thermodynamics. The Conservation of Energy/Mass states the amount of energy/mass is a constant over time. Einstein's Theory of General Relativity defines a time symmetrical black hole and white hole connected by a wormhole.[67] The Second Law of Thermodynamics states disorder increases irreversibly with time and provides an arrow of time direction.

These three laws are listed in the first column of table 4, "The Ultimate Free Lunch Theory versus an Integrated Theory of Everything."

The prevailing cosmological theory, known as the Ultimate Free Lunch,[68] shown in the second column of table 4, states nothing existed before the big bang. That is, the near infinite energy/mass of our universe was created from nothing, or, more precisely, from random energy fluctuations or noise before the big bang. An Integrated Theory of Everything, shown in the third column of table 4, proposes the revolutionary existence of our precursor universe and its super supermassive quark star (matter)/black hole (energy) which created our universe. The Ultimate Free Lunch violates the laws of Conservation of Energy/Mass and Einstein's Theory of General Relativity in deference to the assumed primacy of the Second Law of Thermodynamics. In contrast, An Integrated Theory of Everything satisfies all three laws.

Law	The Ultimate Free Lunch Theory	An Integrated Theory of Everything
Conservation of Energy/Mass	violates	satisfies
Einstein's Theory of General Relativity	violates	satisfies
Second Law of Thermodynamics	satisfies	satisfies

Table 4 The Ultimate Free Lunch Theory versus an Integrated Theory of Everything.

The Ultimate Free Lunch theory violates the Conservation of Energy/Mass because energy/mass was zero before $t = 0$ and our universe's energy/mass was 10^{24} solar masses following $t = 0$. All scientists believe the Conservation of Energy/Mass law is valid during our universe's 13.7 billion years' lifetime. That is, our universe's energy/mass remains constant although the energy/mass type changes. For example, at $t = 0$, all our universe's energy is super force potential energy. As time progresses this energy converts to rest mass,[69] kinetic (translational and rotational), and potential (gravitational, electromagnetic, and nuclear binding) energy/masses. Thus, whereas all scientists believe in the Conservation of Energy/Mass during our universe's lifetime, most scientists believe the Conservation of Energy/Mass is invalid during the transition to our universe at time t = 0. An Integrated Theory of Everything satisfies the Conservation of Energy/Mass because the same energy/mass exists prior to t = 0 in our precursor universe's subset volume as in our universe following t = 0. Our precursor universe's subset volume contained a super supermassive quark star (matter)/black hole (energy) having our universe's energy/mass of 10^{24} solar masses.

Einstein's Theory of General Relativity is time symmetrical about t = 0 and consists of a black hole before t = 0, a white hole after t = 0, and a wormhole or singularity connecting the two. In science fiction, a wormhole is depicted as a tunnel connecting our precursor universe with our universe, where spaceships travel in either time direction between the two universes. In reality, the wormhole is a singularity at the center of a Planck cube as shown in figure 12, "Big bang doughnut-shaped singularity in a Planck cube." The only objects in the wormhole or singularity are super force particles, and they go only in one increasing time direction from the precursor universe to our universe.

The Higgs mechanism was bidirectional. That is, both the matter particle and its associated Higgs force simultaneously condensed or froze out from a super force particle or simultaneously evaporated back to a super force particle. Just prior to t = 0, matter particles and their associated Higgs forces in our precursor universe's super supermassive quark star (matter) evaporated, deflated, and collapsed into super force particles of the super supermassive black hole (energy). The super supermassive quark star's (matter) evaporation to super force particles was the counterpart of the white hole of our universe's big bang. That is, in the white hole or big bang following t = 0, these super force particles condensed into matter particles and their associated Higgs forces.

The Ultimate Free Lunch theory violates the time-symmetrical Einstein's Theory of General Relativity because nothing preceded our universe. In contrast, an Integrated Theory of Everything includes our precursor universe's super supermassive quark star (matter)/black hole (energy), a wormhole or a super force singularity in a Planck cube, a white hole or big bang, and is consistent with Einstein's Theory of General Relativity.

The Ultimate Free Lunch theory satisfies the Second Law of Thermodynamics by assuming the primacy of the latter over the laws of Conservation of Energy/Mass and Einstein's Theory of General Relativity. The Second Law of Thermodynamics states disorder increases irreversibly with time and provides an arrow of time direction. In general, order represents life or available energy, whereas disorder represents death or unavailable energy.

An egg is ordered because it contains life, whereas a broken egg is disordered because it represents death. Egg incubation and hatching result in the birth of a chick, which grows into a chicken, which lays eggs and repeats the life cycle. In contrast, a broken egg represents death. Everyone has seen an egg break, but no one has yet seen a broken egg reconstitute itself. This is an example of increasing disorder with time in accordance with the Second Law of Thermodynamics. A second example is that a live person's molecules are ordered, whereas a dead person's molecules are disordered. The biochemistry of a live person involves carbon-based molecules or nucleic acids (e.g., deoxyribonucleic), proteins, carbohydrates, fats, and other biochemical molecules which interact together to perpetuate and propagate life. In contrast, in the biochemistry of a dead person, molecules stop interacting together, and within hours of death, bacteria and insects decompose the body into disordered molecules.

The super supermassive quark star (matter) to black hole (energy) collapse in a subset volume of our precursor universe reset disorder from maximum to minimum and "resurrected" life via creation of the super force or mother particles. At $t = 0$, our universe had minimum disorder because all particles were superimposed super force particles in the same place, the Planck cube singularity. Life on earth will be eventually destroyed in three billion years by our Milky Way galaxy's collision with the Andromeda galaxy, or in five billion years by our sun's expansion to a red giant. In figure 20, "Disorder versus time," disorder was at a minimum at the start of our universe, or t = 0. After t =0, disorder in our universe continuously rose with time in accordance with the Second Law of Thermodynamics. Minimum disorder at the start of our universe was the reason for the Ultimate Free Lunch's assumption of the primacy of the Second Law of Thermodynamics. That is, if disorder was a minimum at time t = 0, nothing could possibly have preceded the big bang. Thus, the task was to prove minimum disorder at t = 0 was compatible with a precursor universe's super supermassive quark star (matter)/black hole (energy).

Life exists in our solar system. Life propagated and disorder decreased in our solar system, or in a small subset volume of our universe. This decrease in disorder in our solar system occurred without negating the Second Law of Thermodynamics for our entire universe. A molecular cloud's hydrogen and helium gas, dust, and charged particles began ordering themselves at our solar system's creation 4.6 billion years ago. Since our solar system was one of approximately 100 billion Milky Way galaxy stars and our galaxy was one of approximately 100 billion universe galaxies, our solar system's disorder decrease was much less than the increase of our universe's disorder. That is, the decrease of our solar system's disorder was outweighed by the increase of disorder of our universe's other 10^{22} stars.

Similarly, disorder decreased in our precursor universe's subset volume, whereas disorder increased in our entire precursor universe in accordance with the Second Law of Thermodynamics. Just as our sun was one of 10^{22} stars in our universe, our precursor universe's subset volume containing the super supermassive quark star (matter)/black hole (energy) was a small fraction of our total precursor universe's volume. In our precursor universe's subset volume, the super supermassive quark star (matter)/black hole (energy) had two time sequential states: super supermassive quark star (matter) and super supermassive black hole (energy). In the super supermassive quark star (matter) state, disorder increased with time in accordance with the Second Law of Thermodynamics, as shown on the left side of figure 20. However, during the super supermassive quark star (matter) to black hole (energy) evaporation, deflation, and collapse, the maximum disorder super supermassive quark star (matter) switched to the minimum disorder super supermassive black hole (energy). Figure 20 shows this switch from maximum to minimum disorder just prior to the start of our universe at t = 0. In essence, the super supermassive quark star (matter) to black hole (energy) collapse in a subset volume of our precursor universe reset disorder from maximum to minimum and "resurrected" life via creation of the super force or mother particles.

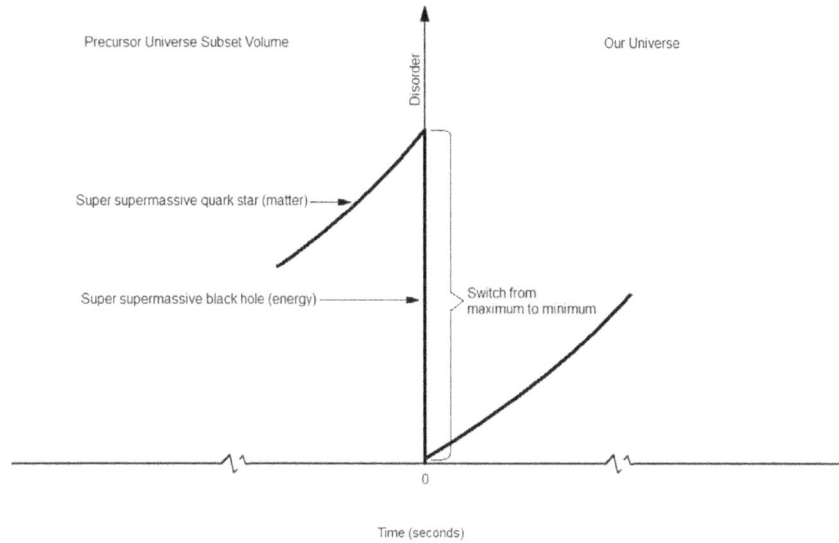

Figure 20. Disorder versus time.

Chapter 14

Super Universe and Our Precursor Universe

Nested within the Super Universe are precursor universes, nested within precursor universes are universes, and nested within universes are galaxies. Three types of stellar objects currently exist in our sky: stars in our Milky Way galaxy, galaxies in our universe, and galaxies in parallel universes within the Super Universe. Stellar objects on the celestial sphere are a superposition of stars in our Milky Way galaxy, galaxies in our universe, and galaxies in parallel universes within the Super Universe.

Nested within the Super Universe are precursor universes, nested within precursor universes are universes, and nested within universes are galaxies. The Super Universe can be represented by a marbles/rising dough/balloons model or an amplified version of the marbles/rising dough/single balloon model for our universe described in chapter 9, "Universe Expansions," and shown in figure 21. Our universe, a parallel universe, and our precursor universe are explicitly shown in figure 21. In the Super Universe model, our universe and a parallel universe are represented by two spherical balloons containing galaxies, our precursor universe by a larger spherical balloon containing our universe and a parallel universe, and the

Super Universe by an even larger spherical balloon containing our precursor universe. The marbles are mixed in rising dough in our universe and the parallel universe's balloons. The rigid marbles represent galaxies in our universe and in the parallel universe. The rigid marbles do not expand. The rising dough represents the space between galaxies, or intergalactic space; space between universes, or interuniversal space; and space between precursor universes, or interprecursor universal space. Intergalactic, interuniversal, and interprecursor universal spaces rise or expand with time. All balloons (our universe, parallel universe, precursor universe, and the Super Universe) expand as intergalactic, interuniversal, and interprecursor universal spaces expand with time.[70] Our universe and a parallel universe are shown nested within our precursor universe in figure 21. Our Milky Way galaxy and another galaxy are shown nested within our universe, and two galaxies are shown nested within a parallel universe. Because of its large relative size, the Super Universe is implicitly outside the precursor universe in figure 21, but is explicitly shown in figure 22, "Super Universe and nested universes."

The first type of stellar object in our sky is the stars in our Milky Way galaxy. Our Milky Way galaxy consists of between 100 billion and 400 billion stars, and its volume can be approximated by a sphere with a radius of 50,000 light-years. The stars are not uniformly distributed in the spherical volume. The Milky Way is actually a thin, flat, spiral disk with a radius of 50,000 light-years and a disk thickness of 1,000 light-years. Our sun is one of the stars in a spiral arm of the galaxy. Milky Way galaxy stars predominantly appear in a narrow overhead band because our spiral galaxy is seen edge-on from earth. Since a fraction of Milky Way stars exist outside the spiral disk and appear in all directions around us, the approximation of a spherical volume with a 50,000 light-year radius is valid.

The second type of stellar object in our sky is the galaxies in our universe. There are approximately 100 billion galaxies distributed uniformly in our universe.[71] There are just as many galaxies overhead, to the right, to the left, or beneath an earth observer. Our universe's galaxies are between our Milky Way galaxy's boundary, or an approximate spherical radius of

50,000 light-years, and our spherical universe's boundary. Our spherical universe has a radius of 46.5 billion light-years, or approximately 440 billion trillion kilometers.[72] Galaxies in our universe may appear as single points of light because of their distance, but sensitive telescopes reveal each of them to contain approximately 100 billion stars.

The third type of stellar object in our sky is the galaxies in parallel universes within the Super Universe. Galaxies in parallel universes within the Super Universe extend from our spherical universe's boundary, or 46.5 billion light-years, to the spherical Super Universe's boundary, or a radius of approximately 10^{50} light-years, or 10^{63} kilometers (see chapter 15, "Cosmological Constant Problem"). The 46.5 billion light-years corresponds to the size of our universe following its expansion over its 13.7 billion years' lifetime. Galaxies exist in parallel universes within our

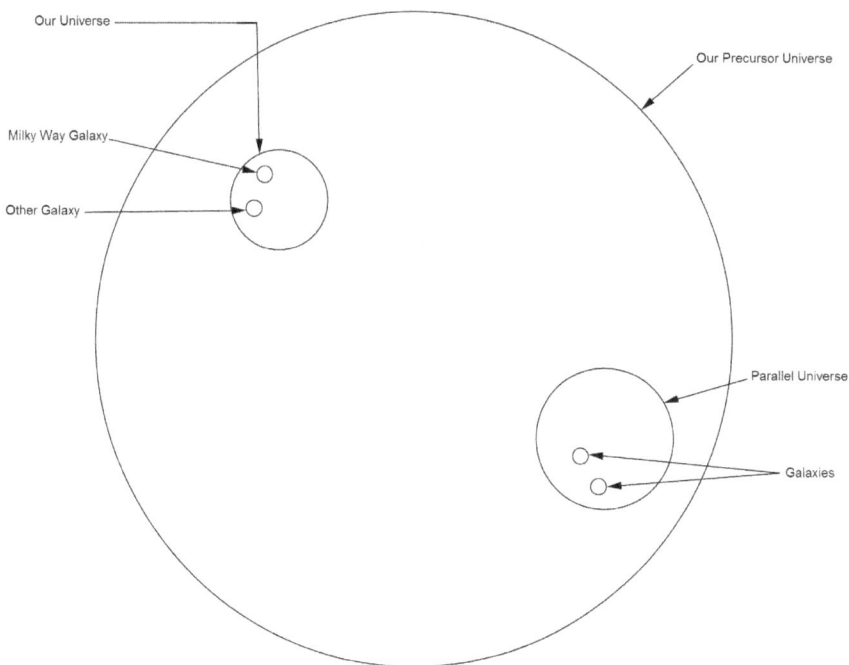

Figure 21. Our precursor universe at the present time.

precursor universe, but also within all parallel universes in all parallel precursor universes in the Super Universe. Therefore, parallel universes exist between our spherical universe's boundary, with a radius of 46.5 billion light-years, and the spherical Super Universe's boundary, with a radius of 10^{50} light-years. Galaxies of parallel universes remain undetected because of their enormous distances from us.

Stellar objects on the celestial sphere are a superposition of stars in our Milky Way galaxy, galaxies in our universe, and galaxies in parallel universes within the Super Universe. All stellar objects appear equidistant from us on the celestial sphere.[73] This is a distorted view because humans lack two basic capabilities of modern telescopes: first, to resolve a stellar object into its components (e.g., a galaxy into its 100 billion component stars); and second, to determine distance or range to the stellar object. Advanced telescopes have the resolution and sensitivity to detect stars within galaxies and to estimate the distance to galaxies using standard candles.[74] Advanced telescopes with higher sensitivity and resolution are required to detect the closest galaxy in the closest parallel universe and its component stars.[75]

Chapter 15

Cosmological Constant Problem

The Super Universe was created first, precursor universes second, universes third, and galaxies fourth. Dark energy and matter were uniformly distributed on a large scale in the Super Universe. Super supermassive quark stars (matter)/black holes (energy) and super super supermassive quark stars (matter)/black holes (energy) formed because matter was concentrated and not uniformly distributed on a small scale in their respective volumes. The observed cosmological constant was 10^{-120} of its expected value and known as the cosmological constant problem.[76] The problem existed because the Super Universe's volume was 10^{120} larger than our universe's volume. Doughnut-shaped super force singularities at the center of Planck cubes existed at the start of the Super Universe, all precursor universes, all universes, and our universe. The expansion rate of galaxies within our universe will eventually become a constant and identical to the expansion rates of our precursor universe and the Super Universe.

The Super Universe was created first, precursor universes second, universes third, and galaxies fourth. Figure 22 shows an example of the Super Universe and nested universes at four sequential big bang times. The Super Universe's big bang occurred at approximately t = -10^{50} years.[77] By an arbitrarily selected time, t = -15 billion years, the Super Universe expanded to a large spherical volume, explicitly shown in figure 22. The Super Universe expanded to a larger spherical volume at t = 0 seconds and to an even larger spherical volume at the present time, t = 13.7 billion years. The Super Universe was not explicitly shown at the two latter times because of its colossal size relative to our universe. During its 10^{50} years' lifetime, the Super Universe continuously expanded and currently is 10^{120} the volume of our universe.[78] At time t = -15 billion years, a super super supermassive black hole (energy)[79] was created in the Super Universe. By time t = 0, the super super supermassive black hole (energy) had transitioned to a white hole and had grown into our precursor universe. At t = 0, a super supermassive black hole (energy) in our precursor universe transitioned to a white hole and created our universe, as described in chapter 12, "Stellar Black Holes."[80] At a time before t = 0, another super supermassive quark star (matter)/black hole (energy) created a parallel universe shown with a finite size at t = 0. In our universe, first-generation stars were created by hydrogen and helium molecular clouds gathering around dark matter clumps. Starting at 200 million years after the start of our universe, these first-generation stars gravitationally collapsed to neutron stars followed by supernova explosions or quark stars (matter) followed by quark-nova explosions, both of which created their associated primeval galaxies. Super super supermassive quark stars (matter)/black holes (energy) were to precursor universes, as super supermassive quark stars (matter)/black holes (energy) were to universes, as supermassive quark stars (matter) were to galaxies. A super super super supermassive black hole (energy) with an energy/mass of 10^{144} solar masses created the Super Universe. This is shown in table 5, "Relationships between stellar black hole types and the Super Universe, precursor universes, universes, and galaxies."

The Super Universe, consisting of parallel universes, was created by time-sequential and concurrent cycles of big bangs through stellar black holes. Three time-sequential cycles are explicitly shown in figure 22 and table 5. First, a super super super supermassive black hole's (energy) big bang created the Super Universe and, at t = -15 billion years, a super super supermassive quark star (matter). Second, the latter's associated super super supermassive black hole's (energy) big bang created our precursor universe and, at t = 0, a super supermassive quark star (matter). Third, the latter's associated super supermassive black hole's (energy) big bang created our universe and, by t = 13.7 billion years, a supermassive quark star (matter) at the center of our Milky Way galaxy. The concurrent or parallel cycles of big bangs through stellar black holes are implicit in figure 22. For example, at t = -15 billion years, a second super super supermassive black hole's (energy) big bang created a second parallel precursor universe and, at t = 0, a super supermassive quark star (matter). The latter's associated super supermassive black hole's (energy) big bang created a parallel universe in the second parallel precursor universe.

Dark energy was uniformly distributed on a large scale in the Super Universe. Universal laws of physics and structure were assumed across the Super Universe. For example, the Super Universe obeyed Conservation of Energy/Mass, contained eleven fundamental matter and force particles, and had a dark energy to total energy/mass ratio identical to our universe. Similar to the chapter 7 model for our universe's dark energy as an expanding balloon filled with black smoke, the model for the Super Universe's dark energy was a much larger expanding balloon filled with black smoke. The expanding balloon represented the Super Universe, and the black smoke represented dark energy. Initially, the balloon, or Super Universe, was small in size, and the black smoke, or dark energy, was very dense. As the balloon expanded, the black smoke density, or dark energy density, decreased. Since the cosmological constant was proportional to dark energy density, as the Super Universe expanded, both dark energy density and the cosmological constant decreased. Dark energy in the Super Universe was a constant and equaled 72.6% of its total energy/mass. The Super Universe's dark energy density, or black smoke density, was uniformly distributed throughout all universes, including the Super Universe, all precursor universes, all universes, and our universe. The right side of figure 22 explicitly shows our universe and a parallel universe at the present time of 13.7 billion years. Both our universe and a parallel universe are nested within our precursor universe, which is nested within the Super Universe. Our precursor universe and the Super Universe are not explicitly shown on the right side of figure 22 because of their large sizes relative to our universe and a parallel universe.

Master Big Bangs through Black Holes in Four Hours

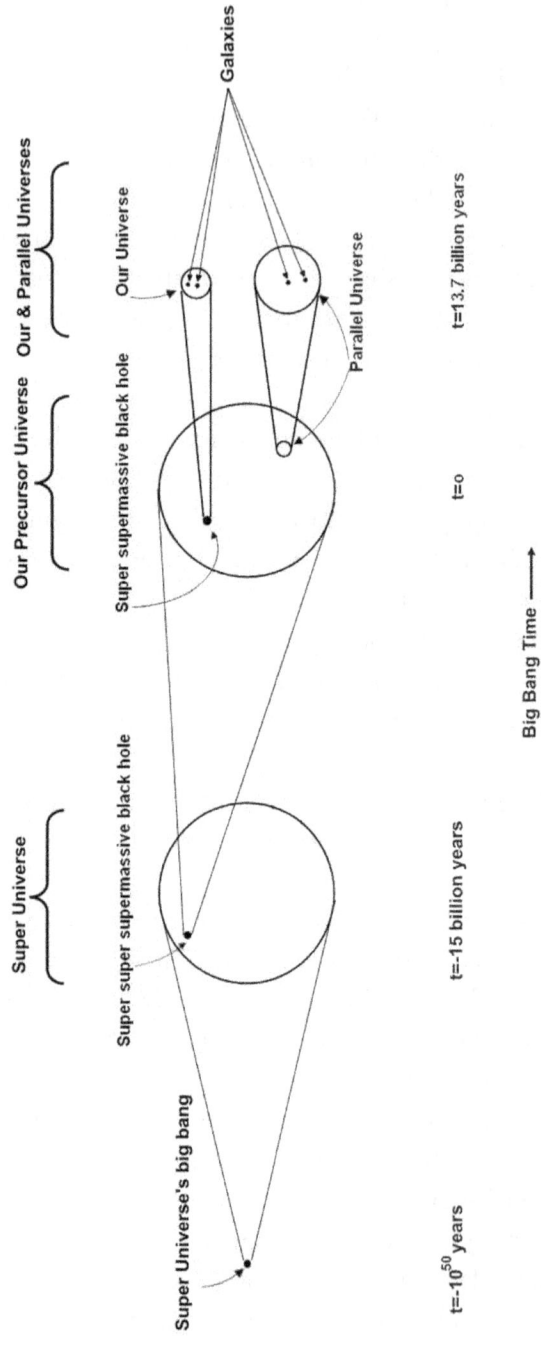

Super Universe and nested universes.

84

Stellar black hole types	Stellar black hole sizes (solar masses)	Super Universe, precursor universes, universes, galaxies
Super super super supermassive black hole (energy)	10^{144}	Super Universe
Super super supermassive quark stars (matter)/black holes (energy)	Much greater than 10^{24}	Precursor universes
Super supermassive quark stars (matter)/black holes (energy)	10^{24}	Universes
Supermassive quark stars (matter)	One million to ten billion	Galaxies

Table 5. Relationships between stellar black hole types and the Super Universe, precursor universes, universes, and galaxies.

Matter is uniformly distributed on a large scale in the Super Universe. At the present time, large scale is defined as a cube with a side equal to approximately 300 million light-years. If our universe is divided into large scale cubes, each cube contains the same amount of matter.[81] At the present time, matter is assumed to be uniformly distributed on the same large scale (i.e., 300 million light-years) in the Super Universe, all precursor universes, all universes, and our universe. If the large scale volume is reduced to the size of a single galaxy, matter is lumpy and not uniformly distributed.

Super supermassive quark stars (matter)/black holes (energy) and super super supermassive quark stars (matter)/black holes (energy) formed because matter was concentrated and not uniformly distributed on a small scale in their respective volumes. In our precursor universe, at approximately time t = 0, or 13.7 billion years ago, subset volumes formed super supermassive quark stars (matter)/black holes (energy), which transitioned to white holes of our universe and a parallel universe. In a subset volume of the Super Universe at an approximate time t = -15 billion years, a super super supermassive quark star (matter)/black hole (energy) formed which transitioned to the white hole of our precursor universe. In these small scale volumes, matter overwhelmed dark energy and caused the above two types of quark stars (matter) to accrete and eventually collapse to their associated black holes (energy).

The observed cosmological constant was 10^{-120} of its expected value and known as the cosmological constant problem. The problem existed because the Super Universe's volume was 10^{120} larger than our universe's volume. The Super Universe was much larger and older than our universe. Since spherical volumes were proportional to their radii cubed, the ratio of the Super Universe's radius R_{su} to our universe's radius R_{ou} was $(10^{120})^{1/3}$ or 10^{40}. The Super Universe's radius was $R_{su} = (10^{40})$ (our universe's radius) $= (10^{40})$ (46.5×10^9 light-years), or approximately 10^{50} light-years. Assuming equal expansion rates of the Super Universe and our universe, that is, our universe's radius/our universe's age = Super Universe's radius/Super Universe's age, the Super Universe's age was estimated to be approximately 10^{50} years.[82] Energy/masses for total, atomic matter, dark matter, and dark energy were uniformly distributed on a large scale in the Super Universe, all precursor universes, all universes, and our universe. This meant the total, atomic matter, dark matter, and dark energy energy/masses in the Super Universe were 10^{120} larger than in our universe. The total energy/mass of the Super Universe or equivalently, the energy of the Super Universe's big bang was (10^{120}) (10^{24} solar masses), or 10^{144} solar masses.

Doughnut-shaped super force singularities at the center of Planck cubes existed at the start of the Super Universe, all precursor universes, all universes, and our universe. At the start of the Super Universe, a super super super supermassive black hole (energy) with energy of 10^{144} solar masses existed. At the start of all precursor universes, super super supermassive black holes (energy) with energy greater than the energy of our universe or 10^{24} solar masses existed. At the start of all universes, super supermassive black holes (energy) with energy of 10^{24} solar masses existed.[83]

The expansion rate of galaxies within our universe will eventually become a constant and identical to the expansion rates of our precursor universe and the Super Universe. During our precursor universe's super supermassive quark star (matter) deflationary collapse, each matter particle and its associated Higgs force evaporated to super force energy leaving a spherical volume with a "perfect" vacuum in its wake. A "perfect" vacuum was completely empty space. Since the super supermassive black hole's (energy) near infinite temperature was much greater than the surrounding "perfect" vacuum's temperature of $0°K$, it transitioned to the white hole and initiated our universe's thermodynamic arrow of time. Our universe decelerated for its first 8 billion years and accelerated during the next 6 billion years. Currently, a spherical shell consisting of a "perfect" vacuum exists between the outer boundary of our universe and the inner boundary of our precursor universe. As our universe accelerates, the spherical shell thickness will approach zero. Our universe's acceleration will stop when our universe's outer boundary merges with our precursor universe's inner boundary. Then, the expansion rate of galaxies within our universe will become a constant and identical to the expansion rate of our precursor universe and the Super Universe.[84]

Chapter 16

Black Hole Information Paradox

Any universe object's intrinsic information was defined by the contents and positions of all the object's contiguous Planck cubes. Our precursor universe's super supermassive black hole (energy) doughnut-shaped singularity consisted of three information parameters: mass, charge, and spin. Our precursor universe's near infinite intrinsic information was irreversibly lost in four star collapse stages during the formation of the super supermassive quark star (matter)/black hole (energy) which created our universe. No intrinsic information swallowed by our precursor universe's super supermassive quark star (matter)/black hole (energy) was emitted as Hawking radiation.

In 1975, Stephen Hawking stated information swallowed by a black hole was not emitted as Hawking radiation. In 2004, his position reversed and information swallowed by a black hole was emitted. This ambiguity was known as the black hole information paradox.

Any universe object's (e.g., an encyclopedia's) intrinsic information is defined by the contents and positions of all the object's contiguous Planck cubes. Intrinsic information consists primarily of the unique relative orientation of up quarks, down quarks, and electrons to each other.

That is, an object's intrinsic information consists of the molecular, atomic, nuclear, and fundamental matter particles' structural information.[85] Each up quark, down quark, and electron of an encyclopedia's ink, paper, and binding molecules resides within a specific Planck cube. At the up quark, down quark, and electron Planck cube level, no two encyclopedias have the same intrinsic information, that is, no two encyclopedias are exactly identical. Since an encyclopedia volume contains a near infinite[86] number of Planck cubes, figure 23 shows only the first letter, *E*, in the word *Encyclopedia* on the book's cover. There is almost a googol of Planck cubes in a grain of sand, and both the letter *E* and the encyclopedia are larger than a grain of sand. Shown are Planck cubes which contain up quarks, down quarks, and electrons associated with the ink's molecules. The figure is not to scale because the Planck cubes are much smaller than shown.

Our precursor universe's super supermassive black hole (energy) doughnut-shaped singularity consisted of three information parameters, mass, charge, and spin, in accordance with the no-hair theorem. The singularity is shown in figure 12, "Big bang doughnut-shaped singularity in a Planck cube." In contrast, our universe contains near infinite intrinsic information. Two intrinsic information examples are: each of the world's books, with their ink, paper, and binding molecules with their unique relative orientation of up quarks, down quarks, and electrons; and each of the world's seven billion people with their carbon-based molecules and their unique relative orientation of up quarks, down quarks, and electrons.[87] Although specific objects in our precursor universe's subset volume were unknown, the subset volume's intrinsic information was assumed to be comparable to our universe's near infinite intrinsic information. This is because our precursor universe's subset volume and our universe had identical fundamental matter and force particles and energy/mass. There were thus comparable numbers of up quarks, down quarks, and electrons in our precursor universe's subset volume and in our universe.

Our precursor universe's near infinite intrinsic information was irreversibly lost in four star collapse stages during the formation of the super supermassive quark star (matter)/black hole (energy) which created our universe. The four collapses were molecules to atoms, atoms to protons/neutrons and electrons, protons/neutrons to quarks, and quarks to super force energy. The four star collapse stages, in increasing order of star density, were white dwarf, neutron star, quark star (matter), and black hole (energy). The process of losing structural information irreversibly was illustrated via a thought experiment in which four hypothetically identical objects, such as encyclopedias, were dropped into increasingly dense stars: white dwarf, neutron, quark star (matter), and black hole (energy). In a white dwarf star, the encyclopedia's molecules were collapsed or decomposed into unbounded atomic nuclei and electrons. Molecular

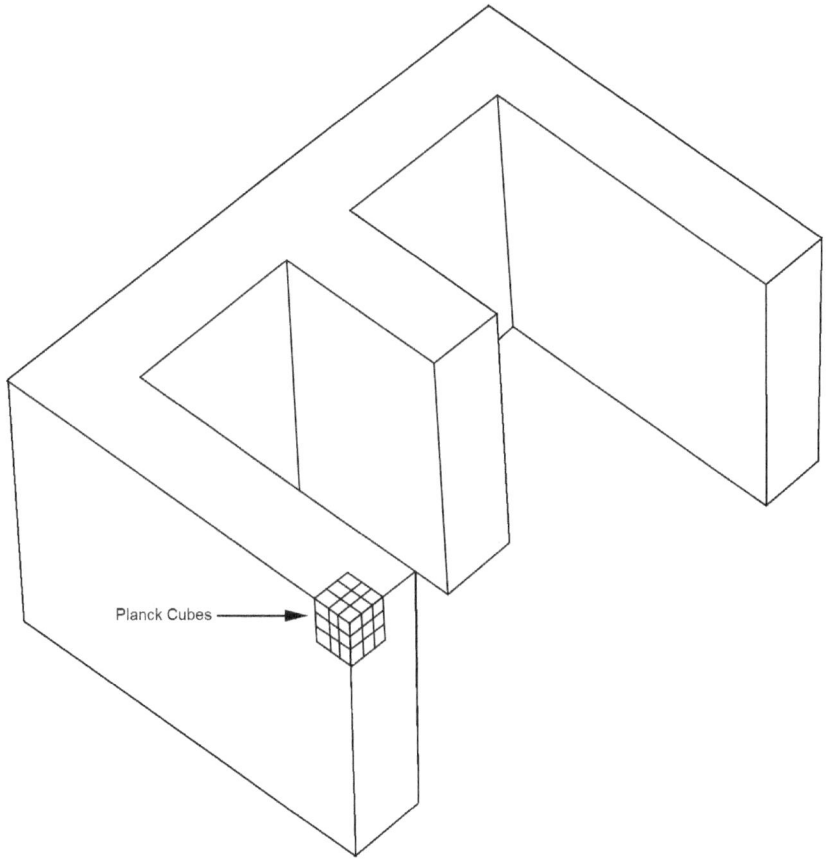

Figure 23. Planck cubes in an encyclopedia s letter E.

structural information was irreversibly lost. A white dwarf star was a dense collapsed star, where electrons prevented further collapse to the next level, a neutron star. In a neutron star, the encyclopedia's atoms were collapsed or decomposed into protons, neutrons, and electrons, and atomic structural information was irreversibly lost. A neutron star was a collapsed star denser than a white dwarf, where neutrons prevented further collapse to the next level, a quark star (matter). In a quark star (matter), the encyclopedia's protons and neutrons were collapsed or decomposed into up quarks and down quarks and nuclear structural information was irreversibly lost. A quark star (matter) was a collapsed star denser than a neutron star, where quarks prevented further collapse to the next level, a black hole (energy) star. In a black hole (energy) star, the encyclopedia's quarks were collapsed or decomposed into super force energy particles and fundamental matter particles' structural information was irreversibly lost.

No intrinsic information swallowed by our precursor universe's super supermassive quark star (matter)/black hole (energy) was emitted as Hawking radiation. Our precursor universe's subset volume originally contained molecular, atomic, nuclear, and fundamental matter information which, over the subset volume's lifetime, was reduced by our precursor universe's super supermassive quark star (matter)/black hole (energy) to the three black hole (energy) parameters of mass, charge, and spin. Our precursor universe's near infinite intrinsic information was lost in four star collapse stages during the formation of the super supermassive quark star (matter)/black hole (energy). The intrinsic information was irreversibly lost and none was emitted as Hawking radiation. Thus, our precursor universe's super supermassive quark star (matter)/black hole (energy) had two stages: destruction of molecular, atomic, and nuclear matter intrinsic information in the quark star (matter) state; and destruction of fundamental matter intrinsic information with the creation of super force energy particles and the "resurrection" of life in the black hole (energy) state.

Chapter 17

Matter/Antimatter

Charge, parity, and time (CPT) violation is the theory which best explains the prominence of matter over antimatter. There are three CPT violation arguments which support each other and this book's conclusions. In the first argument, CPT is violated at the Planck cube scale. In the second argument, highly curved space times such as super supermassive black hole (energy) singularities violate CPT because of incoming matter information disappearance. In the third argument, in the transformation from one quantum state to another, disorder is preserved. At the beginning of matter creation, the Higgs mechanism created an equal number of matter and antimatter particles, but by the end of matter creation, only matter particles existed and remain to the present time. CPT was violated in both our precursor universe's super supermassive black hole (energy) and in its big bang white hole (energy) counterpart. This provided sufficient CPT violations to produce the prominence of matter versus antimatter in our early universe.

CPT violation is the theory which best explains the prominence of matter over antimatter. The prominence of matter versus antimatter in our early universe[88] is one of the major unsolved problems of physics. There are forty-two identified theories of the prominence of matter versus antimatter in our early universe,[89] of which the CPT violation theory provides the best explanation.

There are three CPT violation arguments which support each other and this book's conclusions. In the first CPT violation argument, according to T. D. Lee, CPT is violated at the Planck cube scale.[90] This argument agrees with the conclusion of chapter 12, "Stellar Black Holes." Quantum mechanics is the branch of mechanics based on a quantum or a unit of energy/mass. Each of the five fundamental matter particles (up quark, down quark, electron, zino, and photino) and each of the six force particles (gravitational, electromagnetic, strong, weak, super, and Higgs) exists within a Planck cube or the quantum of energy/mass. In our precursor universe's super supermassive quark star (matter) to black hole (energy) collapse, all matter and force particles in their Planck cubes collapsed to a doughnut-shaped super force singularity at the center of a Planck cube. This is shown in figure 19, "Super supermassive quark star (matter) collapse to a super supermassive black hole (energy)." This doughnut-shaped singularity was much smaller than a Planck cube as shown in figure 12, "Big bang doughnut-shaped singularity in a Planck cube." During the collapse of our precursor universe's super supermassive quark star (matter) to the black hole (energy) super force singularity, quantum mechanics was invalid and CPT was violated because the singularity existed in a volume smaller than a Planck cube or quantum of energy/mass.

In the second CPT violation argument, according to Nick Mavromatos, highly curved space times such as super supermassive black hole (energy) singularities[91] violate CPT because of incoming matter information disappearance.[92] This argument agrees with the conclusion of chapter 16, "Black Hole Information Paradox." That is, our precursor universe's near infinite intrinsic information was irreversibly lost in four star collapse stages during the formation of the super supermassive quark star (matter)/black hole (energy) which created our universe. The intrinsic information was irreversibly lost and none was emitted as Hawking radiation.

In the third CPT violation argument, according to Florian Hulpke, in the transformation from one quantum state to another, disorder is preserved.[93] During the collapse of our precursor universe's super supermassive quark star (matter) to the black hole (energy) super force singularity, quantum mechanics was invalid because the singularity existed in a volume smaller than the Planck cube or quantum of energy/mass. It follows, according to Hulpke's argument, that disorder should *not* be preserved because the transformation from one quantum state to another is invalid. This agrees with the conclusion of chapter 13, "Super Supermassive Quark Star (Matter)/Black Hole (Energy)," where our precursor universe's super supermassive quark star (matter) collapse to a black hole (energy) reset disorder from maximum to minimum and "resurrected" life via creation of the super force or mother particles.

At the beginning of matter creation, the Higgs mechanism created an equal number of matter and antimatter particles, but by the end of matter creation, only matter particles existed and remain to the present time. Figure 6 shows the up quark Higgs mechanism in two dimensions. The up quark Higgs mechanism is actually three dimensional, as shown in figure 24, "Up quark Higgs mechanism in three dimensions." The three-dimensional up quark Higgs mechanism resembles a rolling ball on a Mexican sombrero. The Z axis represents energy density or energy per cubic meter. The amount of energy available for condensation is peak energy density multiplied by our universe's volume at the matter particle's condensation time, that is, Energy = (Energy/unit volume)(volume). The X axis represents one Higgs force particle's energy associated with an up quark matter particle. Similarly, the Y axis represents one Higgs force particle's energy associated with an anti-up quark matter particle.

A particle can be in either of two states, super force or a condensed matter particle and its associated Higgs force. Each of the two states corresponds to a unique ball position on the Mexican sombrero. When the ball is at its peak position at the top of the sombrero at $x = 0$, $y = 0$, $z = 2$ of figure 24, only super force particles exist because condensation to up quarks and their associated Higgs forces has not yet occurred. When the ball rolls down to the position shown in figure 24, or $x = -2$, $y = 0$, $z = 1.5$, super force particles have condensed to up quarks and their associated Higgs forces. The Higgs force energy associated with a single up quark is the x coordinate of the figure 24 ball position, that is $x = -2$. The energy density on the Z axis between 1.5 and 2.0 condensed into up quark matter particles and the energy density between 0.0 and 1.5 condensed into their associated Higgs force particles.

The prominence of up quarks over anti-up quarks occurred as follows. The ball initially moved from its peak position down the Mexican hat equidistant from the X and Y axes. That is, super force particles initially condensed equally into an up quark, its associated Higgs force, an anti-up quark, and its associated Higgs force. The four particles annihilated and evaporated back to super force energy as the ball returned to its peak position. During the second condensation/evaporation cycle, the ball moved down the Mexican hat closer to the X axis than the Y axis and then back to its peak position. After n of these condensation/evaporation cycles, the ball eventually moved to the figure 24 ball position along the X axis, where the super force condensed completely to the up quark and its associated Higgs force and none to the anti-up quark and its associated Higgs force. The figure 24 inset represents the up quark Higgs mechanism following annihilation of anti-up quarks and is identical to figure 6, Up quark Higgs mechanism.

Super force condensations occurred for all five matter particles and produced five associated Higgs forces. There were five unique Higgs mechanisms having the same generic shape of figure 24. However, each matter particle had a different peak energy density and associated Higgs force particle's energy. That is, each of the five Higgs mechanisms had a unique peak position on the Z axis and a unique figure 24 ball position on the X axis following matter condensation. Each of the five matter particle condensations occurred at a different time or temperature during matter creation. The heaviest matter particle (zino or photino) condensed first and the lightest matter particle (electron) condensed last.

CPT was violated in both our precursor universe's super supermassive black hole (energy) and in its big bang white hole (energy) counterpart. Because black and white holes were symmetrical according to Einstein's Theory of Relativity, CPT was violated in both our precursor universe's super supermassive black hole (energy) and in its big bang white hole (energy) counterpart.[94] Each matter particle's evaporation from the super supermassive quark star (matter) to a super force particle in the super supermassive black hole (energy) and each super force particle's condensation from the white hole (energy) to a matter particle in the hot particle soup violated CPT. This provided sufficient CPT violations to produce the prominence of matter versus antimatter in our early universe which exists to the present time.

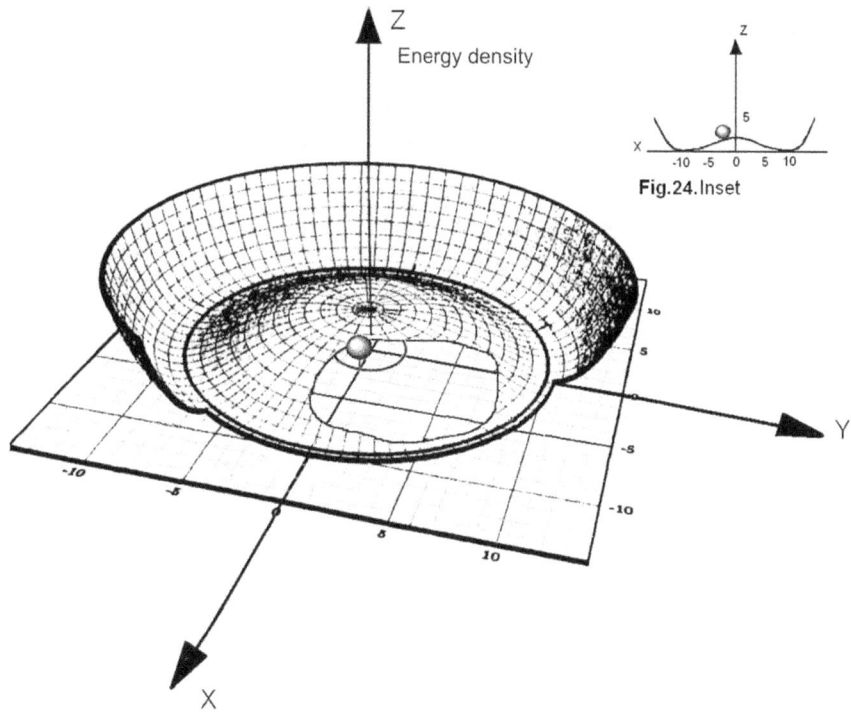

Figure 24. Up quark Higgs mechanism in three dimensions.

Chapter 18

Quantum Gravity Theory

Quantum gravity, string theory, and an Integrated TOE are one because they unify all known physical phenomena from the infinitely small or Planck cube scale (quantum mechanics) to the infinitely large or Super Universe scale (Einstein's General Relativity). Quantum gravity is an evolving theory and one of several quantum gravity theories is string theory.

String theory unifies quantum mechanics with Einstein's General Relativity. All five matter particles (up quark, down quark, electron, zino, and photino) and six force particles (gravitational, electromagnetic, strong, weak, super, and Higgs) exist as Planck cube sized strings with hills' and valleys' amplitude displacement and frequency as discussed in chapter 8, "String Theory," and shown in figure 9, "Particle strings." Since the Planck cube is the quantum or unit of matter particle, force particle, and space, its actions are governed by quantum mechanics.

In contrast, extremely massive and dense bodies such as collapsed stars of the infinitely large Super Universe were governed by Einstein's law of General Relativity. Collapsed stars included white dwarfs, neutron stars, supermassive quark stars (matter) at the centers of all our universe's

100 billion galaxies, super supermassive quark stars (matter)/black holes (energy) which created our universe and parallel universes, and super super supermassive quark stars (matter)/black holes (energy) which created precursor universes. These collapsed stars were discussed in chapter 12, "Stellar Black Holes." and chapter 16, "Black Hole Information Paradox." A collapsed star such as our precursor universe's super supermassive quark star (matter)/black hole (energy) created the doughnut-shaped super force singularity shown in figure 12, "Big bang doughnut-shaped singularity in a Planck cube." This singularity consisted of superimposed super force strings containing all the energy/mass of our universe, or 10^{24} solar masses.

String theory defined all fundamental matter and force particle as strings in Planck cubes. Any Super Universe object was defined by a volume of contiguous Planck cubes each containing a matter or force particle string. Doughnut-shaped super force singularities at the center of Planck cubes existed at the start of the Super Universe, all precursor universes, all universes, and our universe. Thus, string theory unified quantum mechanics of the infinitely small at the Planck cube scale (e.g., fundamental matter and force particles) with Einstein's General Relativity of the infinitely large at the Super Universe scale (e.g., the super super super supermassive black hole (energy) or doughnut-shaped super force strings singularity which created the Super Universe).

Quantum gravity, string theory, and an Integrated TOE which incorporates string theory as described in previous chapters are one because they unify all known physical phenomena from the infinitely small or Planck cube scale (quantum mechanics) to the infinitely large or Super Universe scale (Einstein's General Relativity).

Chapter 19

Systems Engineering

A top down, iterative, systems engineering technique was used to implement an Integrated TOE. Four systems engineering amplification examples were string theory, inflation, the cosmological constant problem, and the Super Universe. Twenty independent existing theories were replaced by an Integrated TOE consisting of twenty interrelated amplified theories.

A top down, iterative, systems engineering technique was used to implement an Integrated TOE. The foundations of an Integrated TOE were twenty independent existing theories, including string, matter and force particle creation, Higgs forces, dark matter, dark energy, universe expansions, Super Universe, stellar black holes, cosmological constant problem, black hole information paradox, matter/antimatter, and quantum gravity.[95] The twenty independent existing theories were developed by physicists primarily for internal integrity with minor emphasis on integration with interrelated theories. The premise of an Integrated TOE is without sacrificing their integrities; these twenty independent existing theories were replaced by twenty interrelated amplified theories.[96]

A top down, iterative, systems engineering technique selectively amplified each independent existing theory without sacrificing the independent existing theory's integrity. Systems engineering was an interdisciplinary field that integrated complex systems such as a radar,[97] a space station, or an Integrated TOE. If each independent existing theory was represented by a jigsaw puzzle piece, the twenty independent existing jigsaw pieces did not fit together because of their independence. Five key independent existing theories or jigsaw puzzle pieces were identified as Higgs forces, particle creation, dark energy, inflation, and string. As shown in figure 25, these five key pieces did not fit snugly together in a jigsaw puzzle. The systems engineering technique iteratively amplified each of the independent existing theories, or, equivalently, iteratively amplified the size and shape of each of the individual jigsaw puzzle pieces. Following are four systems engineering amplification examples.

The first systems engineering example was Brian Greene's string theory, described in chapter 8, "String Theory." A Planck cube sized string's energy/mass was a function of its hill's and valley's amplitude displacement and frequency.[98] Hills' and valleys' amplitude displacement and frequency defined the energy/mass of five individual matter particles (up quark, down quark, electron, zino, and photino) and six individual force particles (gravitational, electromagnetic, strong, weak, super, and Higgs). Greene's definition was sufficient to specify energy/mass of eleven individual matter and force particles, but inadequate to specify the near infinite energy of the doughnut-shaped super force singularity at the start of our universe. Energy/mass definition was amplified in an Integrated TOE to be a function of three, not two string parameters. Energy/mass was primarily a function inversely proportional to string diameter and secondarily proportional to its hill's and valley's amplitude displacement and frequency. Diameter defined the particle's basic energy/mass whereas the hill's and valley's amplitude displacement and frequency fine tuned it. A string with a Planck length diameter and no hill's and valley's amplitude displacement and frequency defined an individual particle with zero energy such as the gravitational force of figure 9(a). A string with a Planck length diameter and having hill's and valley's amplitude displacement and frequency defined a nonzero energy/mass individual particle such as the up quark of figure 9(b).

Since energy/mass was inversely proportional to string diameter, a string diameter much smaller than a Planck length diameter defined the near infinite energy of our universe's doughnut-shaped super force singularity of figure 12. The latter was a singularity of superimposed super force strings whose energy equaled our universe's energy/mass of 10^{54} kilograms. As described in chapter 8, "String Theory," and figure 11, "Three springs potential energy model," diameter defined the particle's basic energy level. The smaller the string diameter, the greater was its potential energy in the compressed springs.

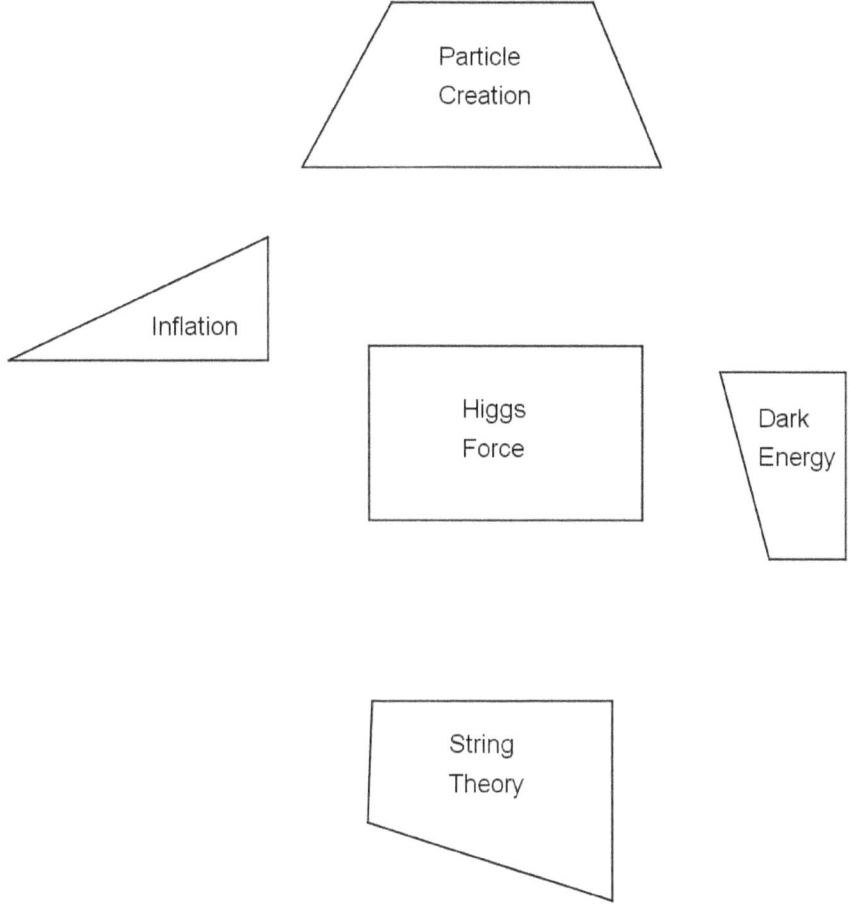

Figure 25. Five key pieces of the Theory of Everything jigsaw puzzle.

The second systems engineering example was Alan Guth's inflation theory. His inflation theory was amplified so that the start of inflation was time synchronous with both the start of matter creation and the one to seven Planck cube energy to matter expansion. In an Integrated TOE, inflation with an exponential inflation factor of 10^{36} began when the universe was the size of a Planck cube. This was less than Guth's exponential inflation factor (10^{49}) but greater than the minimum required (10^{26}).[99]

The third systems engineering example was Paul Steinhardt's cosmological constant problem theory. According to Steinhardt, the cosmological constant problem existed because the universe was older than expected due to precursor cyclical universes.[100] His precursor cyclical universes were amplified to nested precursor universes. Precursor cyclical universes were special cases of nested precursor universes where our precursor universe's super supermassive quark star (matter)/black hole (energy) subset volume equaled our total precursor universe's volume. In Steinhardt's theory, our entire precursor universe collapsed to a black hole. In an Integrated TOE, only a small subset volume of our precursor universe collapsed. In an Integrated TOE, the energy/mass of our precursor universe's super supermassive quark star (matter)/black hole (energy) transitioned to the white hole (big bang) via the Law of Conservation of Energy/Mass. The dark energy coupled to our universe was a fraction of our precursor universe's total dark energy. The fraction equaled our precursor universe's subset volume divided by the total precursor universe's volume.

The fourth systems engineering example was Charles Lineweaver's Extended Ultimate Free Lunch theory, which argued against a Super Universe. His energy density versus time, shown in Figure 26(a),[101] was an extension of the Ultimate Free Lunch theory discussed in chapter 13, "Super Supermassive Quark Star (Matter)/Black Hole (Energy)." Specifically, figure 26(a) is the Ultimate Free Lunch theory extended by addition of the inflation/reheating negative spike. The Integrated TOE is shown in figure 26(b). In the Extended Ultimate Free Lunch theory, the near infinite energy/mass of our universe was created from nothing, or, more precisely, from random energy fluctuations or noise before the start of our universe. This is shown in figure 26(a), where our universe's energy density was zero before t = 0 and instantaneously rose to a near infinite energy density at t = 0. Shortly after t = 0, the extended theory shows a negative spike where our universe's energy density decreases during inflation and increases during reheating or the creation of matter and force particles. Reheating was completed after 10^{-33} seconds. The Extended Ultimate Free Lunch theory violated both the Law of Conservation of Energy/Mass and Einstein's Theory of General Relativity. It also violated basic physics principles of simplicity and symmetry.

Figure 26(b) shows the Integrated TOE's energy density versus time. Since inflation was concurrent with particle creation, the negative inflation/reheating spike was eliminated. The energy/mass of our precursor universe's super supermassive quark star (matter)/black hole (energy) equaled the white hole's energy/mass via the Law of Conservation of Energy/Mass. The super supermassive black hole to white hole transition was symmetrical between -10^{-33} and 10^{-33} seconds and satisfied Einstein's Theory of Relativity. Energy density was symmetrical between -10^{-33} and 10^{-33} seconds because at -10^{-33} seconds, the super supermassive quark star (matter) consisted of a hot particle soup with a radius of 8 meters identical to the hot particle soup of our universe at 10^{-33} seconds. An Integrated TOE also satisfied basic physics principles of simplicity and symmetry.

Twenty independent existing theories were replaced by an Integrated TOE consisting of twenty interrelated amplified theories. After three years and seventy-five research article iterations, the shape of the five jigsaw puzzle pieces representing the five key interrelated amplified theories became a better fit, as shown in the center of figure 27, "An Integrated Theory of Everything jigsaw puzzle." The five independent existing theories are shown by the five unshaded jigsaw puzzle pieces. The shaded areas surrounding the five unshaded jigsaw puzzle pieces represent the five interrelated amplified requirements. The unshaded or independent existing Higgs force jigsaw puzzle piece was amplified by the shaded area to provide interrelated amplified boundaries with the four other jigsaw puzzle pieces (particle creation, dark energy, string theory, and inflation).

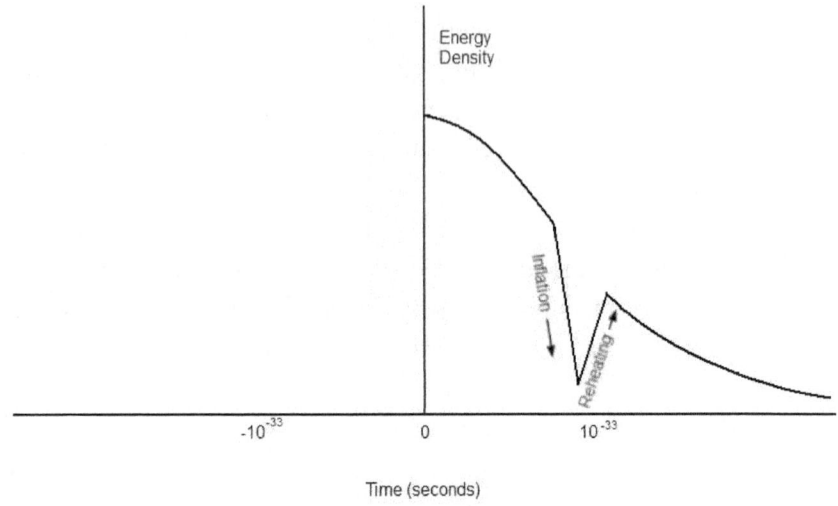

(a) Extended Ultimate Free Lunch Theory

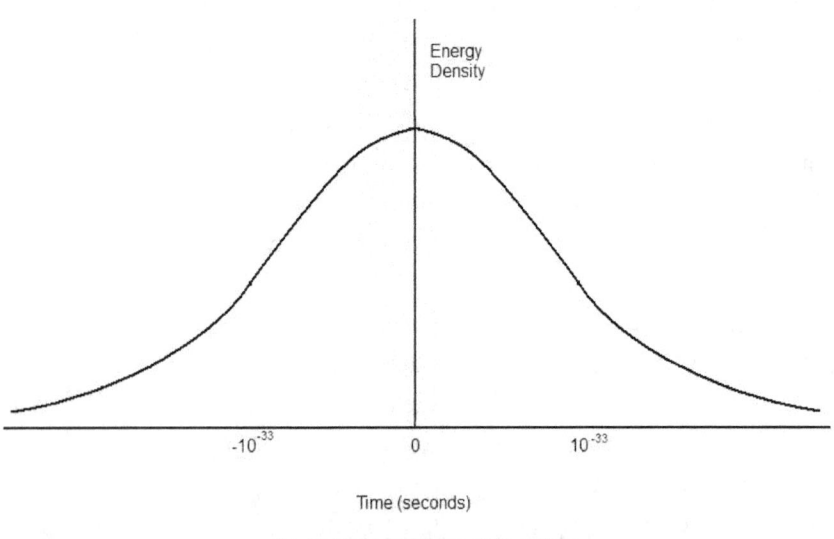

(b) An Integrated Theory of Everything

Figure 26. Extended Ultimate Free Lunch and Integrated Theory of Everything.

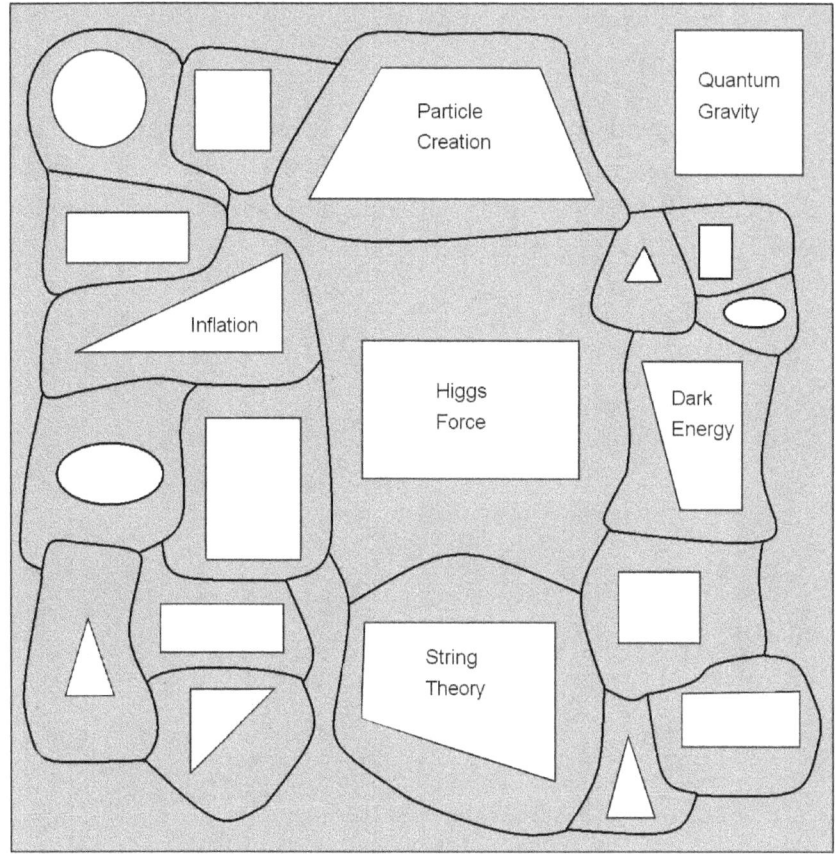

Figure 27. An Integrated Theory of Everything jigsaw puzzle.

After a total of six years and 150 research article iterations, the number of jigsaw puzzle pieces representing independent existing theories expanded to twenty. The twenty jigsaw puzzle pieces were selectively amplified to provide twenty snuggly fitting interrelated amplified theories. The quantum gravity jigsaw piece was amplified to provide interrelated amplified boundaries with the nineteen other jigsaw puzzle pieces. This is because, from the chapter 18, "Quantum Gravity," conclusion, quantum gravity, string theory, and an Integrated TOE unified all known physical phenomena from the infinitely small or Planck cube scale (quantum mechanics) to the infinitely large or Super Universe scale (Einstein's General Relativity).

The twenty independent existing theories were replaced by an Integrated TOE consisting of twenty interrelated amplified theories and summarized by table IV, "Primary interrelationships between twenty amplified theories," in the appendix A research article.

Chapter 20

Conclusions

Everything in our universe is one.

Every animate or inanimate object that exists or existed in our universe is your cousin. I am your cousin. Barack Obama is your cousin. Abraham Lincoln is your cousin. Dinosaurs are your cousins. Poison ivy is your cousin. Mount Everest is your cousin. The supermassive quark star (matter) at the center of our Milky Way galaxy is your cousin. All galaxies in our universe are your cousins. Dark matter is your cousin. Dark energy is your cousin.

At the start of our universe at time = 0, the fundamental building blocks of your body's molecules—your up quarks, down quarks, and electrons—were super force or mother particles. As were my up quarks, down quarks, and electrons, and those of Barack Obama, Abraham Lincoln, dinosaurs, poison ivy, Mount Everest, the supermassive quark star (matter) at the center of our Milky Way galaxy, and all galaxies in our universe. Dark matter and dark energy were also super force or mother particles. At the start of our universe 13.7 billion years ago, all those near infinite compressible super force particles were stacked one atop of another in a doughnut-shaped super force singularity at the center of a single Planck cube.

Everything in our universe is a child of the super force. Everything in our universe is one. We are the Super Force!

Endnotes

[1.] The infinitely small Planck cube scale is very small, finite, and not zero. The infinitely large Super Universe scale is very large, finite, and not infinite.

[2.] The five fundamental permanent matter particles discussed in this book are the most important matter particles. However, in an Integrated TOE there are sixteen Standard Model particles, consisting of twelve fundamental matter particles and four fundamental force particles (one of the four force particles, the W/Z bosons, is actually a transient matter particle); sixteen supersymmetric particles, consisting of four fundamental matter particles and twelve fundamental force particles; thirty-two antiparticles; sixty-four associated supersymmetric Higgs particles; and a super force or mother particle, for a total of 129 particles.

Ignoring antiparticles, there are seventeen fundamental matter particles: six quarks, six leptons, the W/Z boson(s), and four supersymmetric matter particles. The six quarks are top quark, bottom quark, charm quark, strange quark, up quark, and down quark. The six leptons are tau, muon, electron, tau-neutrino, muon-neutrino, and electron-neutrino. The four supersymmetric matter particles are gravitino, gluino, wino/zinos, and photino. See appendix A, "An Integrated Theory of Everything Research Article," table I, "Proposed Standard Model/supersymmetric particle symbols."

[3.] Protons and neutrons are scientifically known as baryons, which are subatomic particles consisting of three quarks.

4. North American Aviation built the GPS satellites. My company, IBM Federal Systems, was responsible for tracking GPS satellites to calculate satellite position data (ephemeris) and transmitting ephemeris data and control commands back to the GPS satellites. Ephemeris data and radio frequency signals were subsequently transmitted from the GPS satellites to users for the latter to calculate distances to satellites and user positions.

5. A person consists of the following biochemical molecules: nucleic acids (e.g., deoxyribonucleic), proteins, carbohydrates, fats, and others.

6. The zino and photino are nonatomic fundamental matter particles scientifically known as nonbaryonic matter. A prime candidate for dark matter is the neutralino, which is an amalgam of the zino, photino, and possibly other particles, including Higgs matter particles or Higgsinos.

7. Additional astronomical observational evidence included the motion of galaxies within galaxy clusters and gravitational lensing. Gravitational lensing was the bending of electromagnetic radiation (e.g., visible light) by intermediate galaxies of radiation originating from more distant source galaxies.

8. Our universe is scientifically described as baryonic (atomic) matter (4.6%), cold dark matter (22.8%), and dark energy (72.6%). Since the mass of electrons is much smaller than up and down quarks, atomic matter is primarily baryonic matter.

9. Direct detection includes the Cryogenic Dark Matter Search (CDMS) at the Soudan mine in Minnesota; the Dark Matter Experiment Using Argon Pulse (DEAP) shape discrimination at the Sudbury Neutrino Observatory (SNO) in Sudbury, Ontario; and the Dark Matter/Large Sodium Iodide Bulk for Rare Processes (DAMA/LIBRA) in the Gran Sasso Laboratory in Italy.

Indirect detection uses gamma and cosmic ray radiation from dark matter annihilation. Gamma rays are high-energy electromagnetic radiation. Cosmic rays are primarily high-energy protons (hydrogen nuclei) and helium nuclei. The function of the Alpha Magnetic Spectrometer (AMS) is to search for dark matter via detection of background positrons, antiprotons, or gamma rays. The

Endnotes

AMS spectrometer was launched by the Space Shuttle Endeavour in 2011 and mounted on the International Space Station.

Collider detection includes production and detection of dark matter by accelerators such as the Large Hadron Collider (LHC) in Geneva, Switzerland.

[10.] This process was scientifically known as recombination.

[11.] A googolth is newly defined by this book as a word meaning a one divided by a googol.

[12.] The binding force between protons and neutrons in a nucleus is scientifically known as the residual strong force.

[13.] For an extremely massive object, such as a stellar black hole, Newton's Law is replaced by Einstein's Theory of General Relativity. For all other cases, Newton's Law is still an excellent approximation of the gravitational force.

[14.] P. K. Kaiser, http://www.yorku.ca/eye/spectru.htm.

[15.] Frequency is the velocity of light divided by wavelength. Electromagnetic radiation energy is defined as $E = hf$, where E is energy, h is Planck's constant, and f is frequency.

[16.] Strong force theory is scientifically known as Quantum Chromodynamics (QCD). If the separation between quarks is less than a proton radius or 10^{-15} meters, the strong force has a property known as asymptotic freedom. If the separation between quarks is greater than a proton radius, the strong force has a property known as confinement.

[17.] α_s is scientifically known as the running or nonlinear coupling constant.

[18.] This is scientifically known as beta minus decay because high-energy electrons, or beta radiation, are produced. In beta minus decay, the down quark transforms into an up quark and a W^- boson or force particle. W^- is one of three types of

W/Z bosons or weak forces known as W⁻, W⁺, and Z^0. Within 10^{-25} seconds, the W⁻ boson transforms into an electron and an anti-electron neutrino. W/Z bosons are transient matter particles as opposed to permanent matter particles such as up quarks, down quarks, and electrons. See appendix A, "An Integrated Theory of Everything Research Article," Superpartner and quark decays/Neutrino oscillations.

[19.] This is scientifically known as beta plus decay.

[20.] The relative strengths of the gravitational and electromagnetic forces were caused by propagation factor dilution between gravitational force activation and electromagnetic/weak force creation/activation in the early universe. See appendix A, "An Integrated Theory of Everything Research Article," Relative strengths of forces/Hierarchy problem.

[21.] The six fundamental force particles discussed in this book are the most important force particles. However, in an Integrated TOE there are sixteen Standard Model particles, consisting of twelve fundamental matter particles and four fundamental force particles (one of the four force particles, the W/Z bosons, is actually a transient matter particle); sixteen supersymmetric particles, consisting of four fundamental matter particles and twelve fundamental force particles; thirty-two anti-particles; sixty-four associated supersymmetric Higgs particles; and a super force or mother particle, for a total of 129 particles.

Ignoring antiparticles, the fifteen force particles are three Standard Model force particles, gravitation (graviton), electromagnetic (photon), and strong (gluon), and twelve supersymmetric force particles: top squark, bottom squark, stau, charm squark, strange squark, smuon, tau-sneutrino, down squark, up squark, selectron, muon-sneutrino, and electron-sneutrino. This book describes five types of Higgs forces, one for each of the five fundamental matter particles. However, there are actually seventeen supersymmetric Higgs forces associated with each of the seventeen Standard Model and supersymmetric matter particles. Similarly, this book describes five types of super forces, one for each of the five fundamental matter particles. However, there are actually seventeen types of super forces associated with each of the seventeen Standard Model and supersymmetric matter particles.

See appendix A, "An Integrated Theory of Everything Research Article," table I, "Proposed Standard Model/supersymmetric particle symbols."

[22.] The Higgs mechanism is known scientifically as spontaneous symmetry breaking.

[23.] Each of five fundamental matter particles has an associated supersymmetric Higgs force particle, and each of four fundamental force particles has an associated supersymmetric Higgs matter particle. Higgs matter particles are scientifically known as Higgsinos. In this book, nine different Higgs particles are mentioned, but only five are analyzed. For analyses of Higgsinos, see appendix A, "An Integrated Theory of Everything Research Article," Supersymmetric Higgs particles.

[24.] Gravitational force messenger particles are scientifically known as gravitons. See appendix A, "An Integrated Theory of Everything Research Article," Messenger particle operational mechanism.

[25.] Lambda (Λ) is the cosmological constant which is proportional to the vacuum or dark energy density (ρ_Λ) as follows, $\Lambda = (8\pi G/3c^2)\, \rho_\Lambda$, where G is the gravitational constant and c is the speed of light. See appendix A, "An Integrated Theory of Everything Research Article," Dark energy.

Dark matter is described as cold because its velocity is nonrelativistic or far below the speed of light. The Lambda-CDM model is consistent with small fluctuations or anisotropies of the Cosmic Microwave Background (CMB) radiation.

[26.] Since the singularity had a very small but finite volume, the super force particles are described as near infinitely compressible.

[27.] The Supernova Cosmology Project, headed by Saul Perlmutter of the Lawrence Berkeley National Laboratory, published in 1998 measurements of our universe's expansion using standard candles or Type Ia supernovas. Saul Perlmutter shared the 2011 Nobel Prize in Physics for his work on this project. A Type Ia supernova explodes from a standard mass and has a standard candle or brightness. When a white dwarf star accretes mass from a binary star and reaches the threshold or standard Chandrasekhar limit of 1.38 solar masses, it violently collapses to a Type 1a supernova.

Additional independent observations analyzed cosmic microwave background, large scale cosmos structures, and gravitational lensing and corroborated the supernova results.

[28.] This book describes eleven fundamental particles, five matter particles and six force particles, which are the most important matter and force particles. However, in an Integrated TOE, there are sixteen Standard Model particles consisting of twelve fundamental matter particles and four fundamental force particles (one of the four force particles, the W/Z bosons, is actually a transient matter particle); sixteen supersymmetric particles, consisting of four fundamental matter particles and twelve fundamental force particles; thirty-two antiparticles, sixty-four associated supersymmetric Higgs particles; and a super force or mother particle, for a total of 129 particles.

Ignoring antiparticles, there are thirty-two Standard/supersymmetric matter and force particles with thirty-two associated Higgs particles consisting of seventeen Higgs forces and fifteen Higgs matter particles known as Higgsinos. Ignoring the fifteen Higgsinos, there are seventeen types of super force particles which condense into seventeen Standard/supersymmetric matter particles and their seventeen associated Higgs force particles. All these individual particles are strings.

The fundamental matter particles (up quark, down quark, and electron) which interact via the force particles (electromagnetic, weak, and strong) and are described in this book are in the Standard Model of physics. The Standard Model actually contains six quarks (up quark, down quark, charm quark, strange quark, top quark, and bottom quark), six leptons (electron, electron-neutrino, muon, muon-neutrino, tau, tau-neutrino), three force particles (electromagnetic, strong, and weak), and the Higgs force particle. The Higgs is the only undetected Standard Model fundamental particle. Since the Standard Model is not a complete theory of interactions—for example, dark matter particles (i.e., zino and photino) and the gravitational force (graviton) are not included in the Standard Model—supersymmetric models were developed to accommodate physics beyond the Standard Model.

Supersymmetry pairs matter particles with force particles. Each Standard Model particle has a partner yet to be detected. For example, an up quark matter particle has an up squark force particle, and a photon force particle has a photino matter particle. The expectation is these high-energy superpartners and Higgs force particles will be detected in the largest particle accelerator in the world, the Large Hadron Collider. Higgs particles are also supersymmetric. If a standard or supersymmetric particle is a matter particle (e.g., an up quark), its associated Higgs particle is a force particle. If a standard or supersymmetic particle is a force particle (e.g., a graviton), its associated Higgs particle is a Higgs matter particle or Higgsino. See appendix A, "An Integrated Theory of Everything Research Article," Proposed standard/supersymmetric particle symbols and Supersymmetric Higgs particles.

[29.] The range of energy/masses of fundamental strings is from 0 electron volts for the graviton or gravitational force to an estimate of 1,500 billion (giga) electron volts (GeV) for the heaviest supersymmetric particle assumed to be the supersymmetric partner of the gravitational force or gravitino.

[30.] The Planck cube sized beach ball shape is known scientifically as a Calabi-Yau membrane. Each of the eleven matter and force particles are equivalently represented three ways: by a dynamic phantom point particle (zero brane), its unique string (one brane), or its associated Calabi-Yau membrane (two brane). In traditional string theory descriptions, a one brane vibrating string generates a two brane Calabi-Yau membrane over time. This concept was amplified so that a zero brane dynamic phantom point particle over time generates quantized particle positions for both a one brane vibrating string and a two brane Calabi-Yau membrane. See appendix A, "An Integrated Theory of Everything Research Article," String theory.

[31.] Pauli's exclusion principle permits force particles (e.g., super force) to exist within the same Planck cube. In contrast, Pauli's exclusion principle prohibits matter particles from occupying the same Planck cube.

32. The transition from a doughnut to spherical shape is scientifically known as a conifold transition.

33. X bosons (inflatons) or the twelve supersymmetric force particles (six squarks and six sleptons) were the latent heat energy source which caused inflation. See appendix A, "An Integrated Theory of Everything Research Article," Supersymmetric Higgs particles and Universe expansions.

34. Pauli's exclusion principle states that no two matter particles can have identical quantum numbers, which was assumed equivalent to occupying the same Planck cube.

35. B. Greene, *The Fabric of the Cosmos* (New York: Vintage Books, 2005), p. 155.

36. M. Rees, ed., *Universe*. (New York: DK Publishing, 2005), pp. 46-49. Kelvin is a unit of temperature measurement starting at absolute zero. Absolute zero is defined as 0° Kelvin or -460° Fahrenheit. For simplicity in this book, only condensation of matter particles is analyzed because condensation of force particles involves Higgs matter particles, or Higgsinos. See appendix A, "An Integrated Theory of Everything Research Article," Supersymmetric Higgs particles.

37. Near infinite is defined as very large, finite, and not infinite. A near infinite number of fundamental atomic matter particles was estimated based on the assumption that all atoms in the universe were hydrogen atoms and the number of hydrogen atoms equaled 10^{80}. Each hydrogen atom contained a proton and an electron, and each proton contained two up quarks and a down quark. Thus, there were 4×10^{80} fundamental atomic matter particles (two up quarks, one down quark, and one electron per hydrogen atom) in our universe. The number of zinos and photinos in our universe was unknown.

38. A. H. Guth, *The Inflationary Universe* (New York: Perseus Publishing, 1997), p. 185, figure 10.6. Figure 15 in this book initialized the inflationary period start radius at $.8 \times 10^{-35}$ meters with an exponential inflation factor of 10^{36}. Guth's comparable values were 10^{-50} meters and 10^{49}. See appendix A, "An Integrated Theory of Everything Research Article," Particle creation sequence/Inflation.

39. The hot particle soup is scientifically known as a hot quark-gluon plasma.

40. The proton and neutron formation period is scientifically known as the hadron era.

41. The first atom production process was scientifically known as big bang nucleosynthesis. This process produced nuclei other than the proton such as deuterium (the heavier isotope of hydrogen), helium-3, helium-4, lithium-6, and lithium-7.

42. Clumping is scientifically known as agglomeration.

43. First-generation stars are scientifically known as Population III stars.

44. Primeval galaxies are scientifically known as protogalaxies.

45. The process of forming neutral hydrogen and helium is scientifically known as recombination.

46. A quark-nova explosion occurred when a neutron star collapsed to its constituent quarks. An estimated 99% of a neutron's energy/mass was the strong force's binding energy and only 1% was in the rest mass energy of the neutron's one up and two down quarks.

47. Matter between stars is scientifically known as interstellar matter. This matter consists of gas, dust, and cosmic rays. Cosmic rays consist primarily of hydrogen and helium nuclei.

48. Matter between galaxies is scientifically known as intergalactic matter. It is thought to consist of low density filaments of ionized hydrogen (protons) and electrons. It is estimated that 50% of atomic (baryonic) matter in our universe is in these filaments with the remainder in galaxies.

49. Trace amounts of lithium and beryllium were created during helium nuclei formation or big bang nucleosynthesis.

50. These two atom production processes are scientifically known as stellar nucleosynthesis and supernova nucleosynthesis, respectively.

51. This process is scientifically known as the r-process for rapid neutron capture.

52. Molecular clouds are scientifically known as nebulas.

53. Some scientists include sulfur and phosphorus atoms. In a recent experiment, arsenic atoms were substituted for phosphorus atoms and life was maintained. This is being disputed.

54. Water's frequency spectrum is at the far-infrared and submillimeter wavelengths. The Herschel telescope and spacecraft launched in May 2009 were specifically designed to detect water in the universe. The greatest distance from earth where water has been detected by the Effelsburg radio telescope is 11.1 billion light-years, or 2.6 billion light-years after the start of our universe.

55. A phase of matter in which a certain portion consists of charged particles, such as positive protons and negative electrons, is scientifically known as a plasma. The sun, stars, and molecular clouds are examples of plasmas.

56. W. Ryan, W. Pittman, *Noah's Flood: the New Scientific Discoveries about the Event that Changed History* (New York: Simon & Schuster, 1999).

57. Supermassive quark stars (matter) are currently described scientifically as supermassive black holes because supermassive quark stars (matter) are hypothetical and have not yet been accepted by the scientific community.

58. Smaller types of black holes (e.g., micro) were not analyzed.

59. C. Carilli, *Science* **323**, 323 (2009). Galaxy to central black hole mass ratio was 30:1 in the early universe and 700:1 now.

60. Three main types of galaxies are elliptical, spiral, and irregular.

61. The radius of the extremely large super supermassive quark star (matter) which created our universe was much less than 10^{26} meters, or $<< 10^{26}$ meters, and estimated as follows. The Schwarzschild radius which defined the event horizon for a nonrotating quark star (matter) was 4×10^{26} meters for our universe's mass. This was the upper radius limit. The lower radius limit occurred if all the matter particles of the super supermassive quark star (matter) were theoretically in contiguous Planck cubes with a radius of approximately 10^{-8} meters. The estimated super supermassive quark star's (matter) radius was between the upper (4×10^{26} meters) and lower (10^{-8} meters) radius limits or approximately $<< 10^{26}$ meters. See appendix A, "An Integrated Theory of Everything Research Article," Big Bang Detection via Gravity Waves and research article endnote.[154]

62. Asteroids are bodies in our solar system in orbit around the sun consisting of rocks, metals, and ice. Most asteroids reside in the asteroid belt between Mars and Jupiter. The belt is estimated to contain approximately 1.5 million asteroids with a diameter larger than 1 kilometer.

63. Comets are bodies in our solar system in orbit around the sun consisting of ice, dust, and rocky particles. The nucleus of Comet McNaught, the largest comet detected to date, is estimated to be between ten and twenty kilometers across. The hypothetically spherical Oort cloud consists of approximately one trillion comets in our solar system.

64. The energy/mass of our universe is 10^{54} kilograms. Since our sun's mass (one solar mass) is approximately 2×10^{30} kilograms, our universe's energy/mass is (10^{54} kilograms) (1 solar mass)/ (2×10^{30} kilograms) or approximately 10^{24} solar masses.

65. The Hawking temperature of a quark star (matter) with mass M was $T=10^{-7}$ ($M_\odot$/M) K and its lifetime t was approximately 10^{66} $(M/M_\odot)^3$ years, where $M_\odot$ was solar mass, and K was degrees Kelvin. The super supermassive quark star (matter) in our precursor universe (mass = $10^{23} M_\odot$) which created our universe had a temperature of $T=10^{-7}$ ($M_\odot/10^{23}M_\odot$) = 10^{-30} K and a life time of approximately 10^{66} $(10^{23}M_\odot/M_\odot)^3$ years = 10^{135} years.

66. Disorder is scientifically known as entropy in the Second Law of Thermodynamics and is proportional to either the quark star's (matter) event horizon area or the quark star's (matter) volume.

67. A wormhole is scientifically known as an Einstein-Rosen bridge.

68. The term "Ultimate Free Lunch" is attributed to Dr. Alan Guth based on the paper "Is the Universe a Vacuum Fluctuation," by E. Tryon, *Nature* (1973). See A. H. Guth, *The Inflationary Universe* (New York: Perseus Publishing, 1997), chapters 1 and 17.

69. Rest mass energy E is defined as $E = mc^2$ where m is the particle's mass and c is the velocity of light.

70. There is a Hubble's law or a linear relationship between the velocity or red shift of a galaxy within our universe and its time or distance. There are also similar Hubble's laws for universes within our precursor universe and precursor universes within the Super Universe. See appendix A, "An Integrated Theory of Everything Research Article," Cosmological constant problem/Nested universes and figure 10.

71. Matter is currently uniformly distributed on a large scale in our universe, where large scale is defined as a cube with a side equal to approximately 300 million light-years.

72. Since a light-year equals 9.46 trillion kilometers, our universe's radius is (46.5 billion light-years) (9.46 trillion kilometers/light-year) ~ 440 billion trillion kilometers.

73. The celestial sphere is an imaginary sphere upon which all stellar objects are projected. All stellar objects on the celestial sphere appear equidistant to an observer on earth.

74. A standard candle is an astronomical object with a consistent luminosity or brightness, for example, a type 1a supernova.

Endnotes

[75.] Advanced telescopes such as the Hubble Ultra Deep Field can detect galaxies with an age of 13.1 billion years. The James Webb telescope, or the post-Hubble space telescope, will be able to detect galaxies with an age several hundred million years older. Neither can detect galaxies in parallel universes with ages greater than 13.7 billion years. The closest galaxy in the closest parallel universe will be detected in the future by a telescope having detection sensitivity beyond 13.7 billion years.

[76.] The observed value was 2×10^{-10} erg/cm^3 and the expected value was 2×10^{110} erg/cm^3 or a discrepancy of 10^{-120}. S. M. Carroll, http://www.livingreviews.org/lrr-2001-1.

[77.] In our universe's big bang time frame, the Super Universe's big bang occurred at $-10^{50} + 13.7 \times 10^{9}$ years ago, or approximately at -10^{50} years. Eventually our universe's big bang time scale, which started at t = 0, should be replaced by the start of the Super Universe, where t = 0 occurred 10^{50} years ago.

[78.] The continuous expansion of the Super Universe was scientifically known as eternal inflation, that is, it occurred during the entire Super Universe's lifetime. Eternal inflation is different than our universe's inflation which occurred immediately after our universe's big bang and lasted less than 10^{-33} seconds.

[79.] The energy/mass of a super super supermassive quark star (matter)/black hole (energy) was much greater than the energy/mass of the super supermassive quark star (matter)/black hole (energy) which created our universe, or 10^{24} solar masses. This is because the super super supermassive quark star (matter)/black hole (energy) which created our precursor universe spawned multiple parallel universes.

[80.] Figure 22 showed one precursor universe in time sequence between the Super Universe and our universe. However, there could be from 0 to n time sequential precursor universes. For n = 0, the Super Universe was our precursor universe. In general, there were n nested time sequential precursor universes.

[81.] Our universe is a cosmic web consisting of galaxy clusters connected by intergalactic filaments. Approximately 50% of atomic matter is assumed to reside in galaxies and 50% in intergalactic filaments. Galactic, intergalactic, and dark matter are uniformly distributed in our universe on a large scale.

[82.] Our universe's radius R_{ou}/our universe's age A_{ou} = Super Universe's radius R_{su}/Super Universe's age A_{su} or $(R_{ou}/A_{ou}) = R_{su}/A_{su}$ or $A_{su} = (R_{su}/R_{ou})(A_{ou}) = (10^{40})$ (13.7 billion years) ~ 10^{50} years.

[83.] To provide a variety of sizes for a quark star (matter) to its associated black hole (energy) collapse, collapse size was assumed to be a function of two thresholds, energy/mass and energy/mass density. See appendix A, "An Integrated Theory of Everything Research Article," Cosmological constant problem/Nested universes, and endnote [158].

[84.] See appendix A, "An Integrated Theory of Everything Research Article," Cosmological constant problem/Nested universes, and figure 10, Hubble's law.

[85.] The remaining encyclopedia's Planck cubes contain other matter and force particles, for example, zino and photino dark matter particles and gravitational, electromagnetic, strong, weak, super, and Higgs force particles. For simplicity, these other matter and force particles are excluded from this analysis.

[86.] Near infinite is defined as finite but very large. For example, there is almost a googol of Planck cubes in each grain of sand.

[87.] A universe object's (e.g., an encyclopedia's) intrinsic information consists of its molecular, atomic, nuclear, and fundamental matter (e.g., up quark) structure. In contrast, an encyclopedia's extrinsic information consists, for example, of its written words. An encyclopedia and a pile of manure having the encyclopedia's precise dimensions and number of Planck cubes have a similar amount but different intrinsic information. In contrast, the encyclopedia has significant extrinsic information (e.g., its written words) whereas the comparable pile of manure has none.

[88.] The study of matter and antimatter production in our early universe is scientifically known as baryogenesis. The asymmetric production of baryons and antibaryons in the early universe is expressed as the baryon to photon ratio η = 6.1 x 10^{-10}. Baryons are protons and neutrons, the building blocks of atomic nuclei. See appendix A, "An Integrated Theory of Everything Research Article," Baryogenesis.

[89.] Of the forty-two identified baryogenesis theories, the following six are prominent: electroweak, Grand Unified Theory (GUT), quantum gravity, leptogenesis, Affleck-Dine, and CPT violation.

[90.] T. D. Lee, *Selected Papers, 1985-1996* (Amsterdam: Gordon and Breach, 1998), p. 776, p. 787.

[91.] Einstein's general relativity is based on three-dimensional (e.g., Planck cubes) space. If space is represented by a rectangular grid (e.g., Planck squares) on a rubber fabric (i.e., a two-dimensional simplifying visualization), then pinching and pulling down the fabric into a funnel shape at the energy/mass [e.g., super supermassive black hole (energy)] location represents gravity. Highly curved space time is the funnel shape of the super supermassive black hole (energy).

[92.] N. E. Mavromatos, http://arxiv.org/PS_cache/hep-ph/pdf/0504/0504143v1.pdf.

[93.] F. Hulpke et al., *Foundations of Physics*, **36**, 479, 494 (2006).

[94.] The big bang's precursor universe's super supermassive quark star (matter)/black hole (energy) and its white hole (energy)/matter counterpart are detectable via gravity waves. The location of the estimated big bang gravitational waveform is the origin of our universe's big bang and it occurred at the big bang time t = 0, or 13.7 billion years ago. The estimated big bang gravitational waveform consists of a pulse and a decaying step function, both having equal maximum amplitudes. The estimated gravitational energy waveform should be detectable at the big bang's location and time by an advanced Laser Interferometer Gravitational Observatory (LIGO) or Laser Interferometer Space Antenna (LISA).

(See appendix A, "An Integrated Theory of Everything Research Article," Big Bang Detection via Gravity Waves and figure 7).

[95.] For all twenty independent existing theories, see appendix A, "An Integrated Theory of Everything Research Article," table IV, Primary interrelationships between twenty amplified theories.

[96.] Approximately ten additional independent existing theories were analyzed and discarded as incompatible with an Integrated Theory of Everything. For example, quintessence was analyzed and discarded. In quintessence, dynamic dark energy varied with time. Quantum loop gravity was also analyzed and temporarily not included pending further investigation.

[97.] A radar consists of the following integrated complex systems: transmitter, receiver, antenna, command and control, computer, and external interfaces.

[98.] B. Greene, *The Elegant Universe* (New York: Vintage Books, 2000), p. 144. Greene specifies amplitude displacement and wavelength (frequency).

[99.] See appendix A, "An Integrated Theory of Everything Research Article," Particle creation/Inflation. A. H. Guth, *The Inflationary Universe* (New York: Perseus Publishing, 1997), is summarized as follows: The research article initialized the inflationary period start radius at $.8 \times 10^{-35}$ meters with an exponential inflation factor of 10^{36}. Guth's comparable values were 10^{-50} meters and 10^{49}. Liddle and Lyth specify an exponential inflation factor greater than 10^{26}. Amplifying the start of inflation to be time synchronous with the start of matter creation eliminated the problem of reheating or the generation of matter following inflation.

[100.] P. J. Steinhardt, N. Turok, *Endless Universe: Beyond the Big Bang* (New York: Doubleday, 2007), p. 249.

[101.] C. H. Lineweaver, http://ned.ipac.caltech.edu/level5/March03/Lineweaver/Lineweaver4_4.html.

Glossary

Antimatter. Antimatter has the same mass as its associated matter particle but an opposite electrical charge, for example, an anti-up quark and an up quark.

Arrow of time. The Second Law of Thermodynamics states disorder increases irreversibly with time and provides an arrow of time direction.

Asteroids. Asteroids are small solar system bodies consisting of rocks, metals, and ice.

Atom. The atom is the second atomic building block level and consists of protons, neutrons, and electrons.

Atom production processes. Atom production occurs during three astronomical processes: during the creation of our universe (helium nuclei formation), in the formation of large star cores, and during supernova explosions.

Atomic building block levels. There are five atomic building block levels: first (molecule); second (atom); third (proton, neutron, and electron); fourth (up quark, down quark, and electron); and fifth (string).

Atomic fundamental matter particles. Atomic fundamental matter particles consist of the up quark, down quark, and electron.

Atomic number. Atomic number is the number of protons in the atom's nucleus which identifies a chemical element.

Atom's nucleus. The atom's nucleus is the center of an atom and consists of protons and neutrons.

Baryogenesis. Baryogenesis is the prominence of matter over antimatter in the early universe. That prominence remains today.

Baryon. A baryon is an atomic particle consisting of three quarks (e.g., a proton or a neutron).

Bidirectional Higgs mechanism. The bidirectional Higgs mechanism supports condensation and evaporation of matter particles and their associated Higgs forces from and to the super force.

Big bang. The big bang includes the start and early expansion of our universe.

Big bang time. The start of big bang time is t = 0, or 13.7 billion years ago.

Black hole (energy). A black hole (energy) is a stellar black hole consisting entirely of energy.

Black hole information paradox. The black hole information paradox is the ambiguity as to whether information from a black hole is or is not emitted as Hawking radiation.

Building block levels. Matter and force particles are decomposed into five building block levels, as shown in figure 8. The five atomic building block levels are molecule; atom; proton, neutron, and electron; up quark, down quark, and electron; and string.

Carbon-based molecules. Carbon-based molecules are required for life and consist of carbon, hydrogen, oxygen, and nitrogen atoms. Water molecules are also required for life.

Celestial sphere. The celestial sphere is an imaginary sphere upon which all sky objects are projected. All stellar objects on the celestial sphere appear equidistant to an observer on earth.

Chemical element. A chemical element is a type of atom described by its atomic number; for example, hydrogen (H) has atomic number 1 and carbon (C) has atomic number 6.

Comets. Comets are solar system bodies consisting primarily of ice.

Condensations. Super force condensations occur for five matter particles and their five associated Higgs forces during particle creation.

Conservation of Energy/Mass. The Conservation of Energy/Mass is a law of physics which states the amount of energy/mass is a constant over time.

Contiguous Planck cubes. Any Super Universe object is representable by a volume of contiguous Planck cubes. Planck cubes are visualized as near infinitely small, cubic, Lego blocks.

Cosmological constant. The cosmological constant is proportional to dark energy density. As our universe expanded with time, the cosmological constant decreased along with dark energy density. Thus, the cosmological constant is not a constant but a decreasing variable of time.

Cosmological constant problem. The cosmological constant problem is the 10^{-120} discrepancy between its observed and expected values.

Cosmology. Cosmology is the study of the origin, structure, and evolution of our universe.

CPT violation. Charge, parity, time (CPT) violation is the theory which best explains the prominence of matter over antimatter.

Dark energy. Dark energy is the unknown and proposed form of energy to explain our universe's expansion. Dark energy is the sum of Higgs force particles' energies associated with five permanent matter particles (down quark, up quark, electron, zino, and photino).

Dark matter detection. Three dark matter detection techniques are direct, indirect, and collider.

Dark matter. Dark matter consists of two undetected zino and photino fundamental matter particles. Both are nonatomic matter particles.

Deflation. Deflation is the opposite of inflation. Our precursor universe's super supermassive quark star (matter) instantaneously deflated and collapsed to the doughnut-shaped singularity of the super supermassive black hole (energy). The deflation time was estimated to be approximately twice our universe's big bang inflation time.

Disorder. The Second Law of Thermodynamics states disorder increases irreversibly with time in an isolated system and provides an arrow of time direction. In general, order represents life or available energy whereas disorder represents death or unavailable energy.

Down quark. The down quark is one of three fundamental atomic fourth level matter particles (up quark, down quark, and electron) and has a negative fractional (-1/3) charge.

Einstein's equation. Einstein's equation $E = mc^2$ shows the relationship between mass and energy, where E is energy, m is mass, and c is the velocity of light.

Einstein's Theory of General Relativity. Einstein's Theory of General Relativity includes a time-symmetrical black hole and white hole connected by a wormhole.

Electromagnetic force. The electromagnetic force is the attractive force of two opposite polarity electromagnets and the repulsive force of two like polarity electromagnets. Electromagnetic force holds electrons and protons together inside atoms and electrons of atoms together inside molecules.

Electromagnetic radiation types. Electromagnetic radiation types consist of gamma rays, x-rays, ultraviolet rays, visible light rays, infrared rays, radar, frequency modulated (FM) radio waves, television (TV) waves, shortwave radio waves, and amplitude modulated (AM) radio waves.

Electron. The electron is one of three atomic third building block level matter particles (proton, neutron, and electron) and one of three fundamental atomic fourth building block level matter particles (up quark, down quark, and electron). The electron has a negative charge.

Electrostatic Force Law. Coulomb's Electrostatic Law between charged particles at rest has the same form as Newton's Law, and its equation is $F = Cq_1q_2/r^2$, where q_1 and q_2 are two charges, r is the range between charges, and C is Coulomb's constant.

Electroweak theory. Electroweak theory defines the electromagnetic and weak forces to be components of the combined electroweak force.

Energy/mass equivalency. The equivalency of energy (E) and mass (m) is in accordance with Einstein's formula $E = mc^2$, where c is the velocity of light.

Eternal inflation. Eternal inflation is the continuous expansion of the Super Universe during its life time (10^{50} years).

Evaporations. Evaporations occurred for five matter particles and their five associated Higgs forces to the super force during our precursor universe's super supermassive quark star (matter) collapse to its associated super supermassive black hole (energy).

Exponential notation. Exponential notation is a method of describing extreme range of numbers required for the infinitely large and small numbers in the TOE.

Extended Ultimate Free Lunch. The Extended Ultimate Free Lunch theory is an extension of the Ultimate Free Lunch theory which explicitly includes the negative spike of inflation and reheating.

First-generation stars. First-generation stars containing hydrogen and helium atoms were first created 200 million years after the start of our universe at dark matter framework nodes. These first-generation stars contained up to 100 times more hydrogen and helium gas than our sun, had short lives, exploded as supernovas or quark-novas, and created over 100 billion neutron and quark stars (matter) and their associated primeval galaxies.

Force particle. A force particle is one consisting of energy. There are six fundamental force particles: four known (gravitational, electromagnetic, strong, and weak) and two proposed but undetected (super and Higgs).

Four star collapse stages. The four star collapse stages in increasing order of star density are white dwarf, neutron star, quark star (matter), and black hole (energy).

Framework nodes of our universe's galaxies. Over 100 billion clumps of dark matter occurred in our early universe, and these were the framework nodes or three-dimensional locations of our universe's 100 billion galaxies.

Fundamental matter and force particles. The five fundamental matter particles are up quark, down quark, electron, zino, and photino. The six fundamental force particles are gravitational, electromagnetic, strong, weak, super, and Higgs.

Galaxy. A galaxy is a group of approximately 100 billion stars held together by gravity. There are over 100 billion galaxies in our universe.

Googol. A googol is an extremely large number defined by a one followed by 100 zeros.

Googolth. A googolth is a newly defined word meaning a one divided by a googol.

Gravitational force. The gravitational force is the attractive force between masses, such as the earth holding people on its surface and the moon in its orbit.

Gravitational force law. Newton's law states gravitational force is proportional to the product of two masses and inversely proportional to the square of the range between the masses. Newton's equation is $F = Gm_1m_2/r^2$, where m_1 and m_2 are two masses, r is the range between masses, and G is the gravitational constant.

Gravitational lensing. Gravitational lensing is the bending of electromagnetic radiation (e.g., visible light) by intermediate galaxies of radiation originating from more distant source galaxies.

Hierarchy problem. The hierarchy problem is the weakness of the gravitational force relative to the electromagnetic force by a factor of 10^{-42}.

Higgs force. The Higgs force or Higgs boson is a residual super force particle which gives mass to its associated matter particle. The sum of Higgs force energies associated with five matter particles is dark energy.

Higgs mechanism. The process of generating five matter particles (up quark, down quark, electron, zino, and photino) and their five associated Higgs force particles is the Higgs mechanism. The Higgs mechanism resembles a rolling ball on a Mexican sombrero.

Higgsino. A Higgs matter particle is known as a Higgsino.

Highly curved space time. An example of highly curved space time is a super supermassive black hole (energy) singularity. If space is represented by a rectangular grid (e.g., Planck squares) on a rubber fabric (i.e., a two-dimensional simplifying visualization), then pinching and pulling down the fabric into a funnel shape at the energy/mass [e.g., super supermassive black hole (energy)] location represents gravity. Highly curved space time is the funnel shape of the super supermassive black hole (energy).

Hill's and valley's amplitude displacement and frequency. A string is visualized as a thin sticky band wrapped around a Planck cube sized beach ball having surface hill's and valley's amplitude displacement and frequency. The gravitational string shown in figure 9a has no hill's and valley's amplitude displacement and frequency and is a perfect circle. In contrast, the up quark's string in figure 9b is shown as a circle with hill's and valley's amplitude displacement and frequency.

Hot particle soup. Our universe expanded from a singularity to a hot particle soup by the end of inflation. The hot particle soup consisted of a subset of up quark, down quark, electron, zino, and photino matter particles.

Independent existing theories. The foundations of an Integrated TOE are twenty independent existing theories, including string, matter and force particle creation, Higgs forces (Higgs bosons or God particles), dark matter, universe expansions, dark energy, Super Universe (multiverse consisting of parallel universes), stellar black holes, black hole information paradox, and baryogenesis (prominence of matter over antimatter).

Inflation. Inflation was the exponential increase of the size of our universe. Inflation occurred immediately following the big bang singularity and caused the size of our universe to inflate from a sphere with a radius of $.8 \times 10^{-35}$ meters just touching the sides of a Planck cube, to a sphere with a radius of eight meters, or an exponential inflation factor of $8/.8 \times 10^{-35} = 10^{36}$. This occurred in less than a millionth, of a billionth, of a billionth, of a billionth of a second, or 10^{-33} seconds.

Intergalactic space. Intergalactic space is the space between galaxies.

Interrelated amplified theories. The premise of an Integrated TOE is without sacrificing their integrities; twenty independent existing theories are replaced by twenty interrelated amplified theories.

Interuniversal space. Interuniversal space is the space between universes.

Integrated Theory of Everything (Integrated TOE). An Integrated Theory of Everything consists of twenty interrelated amplified theories which replaced twenty independent existing theories.

Intrinsic information. Any universe object's intrinsic information (e.g., an encyclopedia's), is defined by the contents and positions of all the object's contiguous Planck cubes. It consists of the molecular, atomic, nuclear, and fundamental matter particles' structural information.

Isotopes. Isotopes are variants of atoms or elements having different numbers of neutrons. That is, they have the same atomic number (number of protons) but different mass numbers (number of protons and neutrons).

Jigsaw puzzle piece. If each independent existing theory is represented by a jigsaw puzzle piece, they do not fit together because of their independence. Selectively adding new interrelated requirements to the twenty independent existing theories is equivalent to selectively amplifying the sizes and shapes of twenty jigsaw puzzle pieces. This selective amplification process produces twenty snuggly fitting interrelated and amplified jigsaw puzzle pieces for an Integrated TOE.

Kilometer. The kilometer is a metric unit of length equal to 1,000 meters. One kilometer equals .621 miles.

Lambda-Cold Dark Matter (Λ-CDM) model. The Lambda-Cold Dark Matter model is the standard model of cosmology.

Large scale. At the present time, large scale in our universe and the Super Universe is defined as a cube with a side equal to approximately 300 million light-years. A cube of this size placed any where in the Super Universe will contain the same amount of matter.

Light-year. A light-year is a unit of length equal to 9.46 trillion kilometers.

Marbles/rising dough/balloon model. Our universe's nonuniform distribution of matter expansion can be represented by a marbles/rising dough/balloon model. The marbles are mixed in rising dough in an expanding balloon. The rigid marbles represent galaxies which do not expand in size and are nonuniformly distributed in the balloon. The rising dough represents intergalactic space or space between galaxies. The dough or intergalactic space rises or expands with time. The spherical balloon represents our universe, which expands with time along with intergalactic space.

Mass number. Mass number is the number of protons and neutrons in the nucleus.

Matter particle. A matter particle is one consisting of mass. There are five fundamental matter particles: three known atomic matter particles (up quark, down quark, and the electron) and two proposed but undetected dark matter particles (zino and photino).

Messenger particles. Mass is given to a matter particle via its associated Higgs force and gravitational force messenger particles transmitted between matter particles.

Meter. The meter is a metric unit of length equal to 3.28 feet.

Mexican sombrero. The Higgs mechanism resembles a rolling ball on a Mexican sombrero.

Milky Way galaxy. The Milky Way galaxy is a group of approximately 100 billion stars held together by gravity. One of the stars is our sun.

Glossary

Molecular clouds. Molecular clouds consist primarily of molecular hydrogen and are the stellar nurseries where stars are formed.

Molecule. The molecule is the first atomic building block level and consists of atoms.

Multiverse. The Super Universe is a multiverse consisting of parallel universes.

Near infinite. Near infinite is defined as very large but finite.

Nested universes. Our universe is nested within our precursor universe which is nested within the Super Universe.

Neutron. A neutron is an atomic third building block level matter particle consisting of two down quarks and an up quark and having no charge.

Neutron star. A neutron star is a dense star consisting primarily of neutrons. A neutron star is more dense than a white dwarf star but less dense than a quark star (matter).

No-hair theorem. The no-hair theorem states black holes (energy) have only three information parameters: mass, charge, and spin.

Nonuniform distribution of matter. Matter was not uniformly distributed (e.g., it was lumpy) from 30,000 years to the present time. Following 30,000 years, samples any where in our universe did not contain the same number of matter particles.

Nuclear binding energy. Nuclear binding energy is the energy required to split an atom's nucleus into its components.

One to seven Planck cubes energy to matter expansion. The one to seven Planck cubes expansion consisted of six contiguous Planck cubes containing six matter particles, attached to the six faces of our universe's original Planck cube containing super force particles.

Our precursor universe. Our precursor universe preceded our universe in time. Nested within our precursor universe were our universe and parallel universes.

Our universe. Our universe consists of approximately 100 billion galaxies each of which consists of approximately 100 billion stars. Our universe is spherical in shape with an approximate radius of 46.5 billion light-years or 440 billion trillion kilometers.

Parallel universe. A parallel universe exists concurrently in time with our universe but spatially outside our universe's boundary. The Super Universe is a multiverse consisting of parallel universes.

Particle. A particle is either one of five matter particles (e.g., up quark) or one of six force particles (e.g., Higgs force).

Particle accelerator. A particle accelerator uses electromagnetic fields to accelerate and contain charged particles. The largest and most powerful particle accelerator is the Large Hadron Collider in Geneva, Switzerland. It is 27 kilometers in circumference and has 7 tera electron volts or 7 TeV of power.

Particle creation. Particle creation is the condensation of super force particles at nine different temperatures to five matter (up quark, down quark, electron, zino, and photino) and four force (gravitational, electromagnetic, strong, and weak) particles and their nine associated Higgs particles.

Pauli's exclusion principle. Pauli's exclusion principle states matter particles cannot occupy the same quantum state whereas force particles can occupy the same quantum state. This was interpreted as matter particles cannot occupy the same Planck cube whereas force particles can occupy the same Planck cube.

Perfect vacuum. A perfect vacuum is completely empty space.

Periodic table of elements. The periodic table of elements contains known chemical elements, such as hydrogen (H), helium (He), oxygen

(O), nitrogen (N), iron (Fe), silver (Ag), gold (Au), lead (Pb), etc., organized by atomic structure properties.

Planck cube. The Planck cube is the universe's fundamental building block and is the quantum or unit of matter particle, force particle, and space. Each of the five fundamental matter and six force particles exists within a Planck cube. A Planck cube side has a Planck length of 1.6×10^{-35} meters.

Planck length. A Planck length is 1.6×10^{-35} meters and is the length of a Planck cube side.

Photino. The photino is one of the two fundamental dark matter particles. The second is a zino.

Plasma. Plasma is a state of matter in which a certain portion of the particles consists of charged particles such as positive protons and negative electrons.

Precursor universe. Our universe is nested within our precursor universe which is nested within the Super Universe. The Super Universe was created before our precursor universe and our precursor universe was created before our universe.

Primeval galaxy. A primeval galaxy consists of matter and energy ejected into space following a gravitational collapse of a massive star to either a neutron or quark star (matter) and the subsequent supernova or quarknova explosion. Stars are formed from this primeval galaxy ejected matter.

Proton. A proton is an atomic third building block level matter particle consisting of two up quarks and a down quark and having a positive charge

Quantum. A quantum is a unit, for example, a Planck cube is the quantum or unit of matter particle, force particle, and space.

Quantum gravity. Quantum gravity is an evolving theory that unifies quantum mechanics at the infinitely small Planck cube scale to Einstein's

General Relativity at the infinitely large Super Universe scale. String theory is one of several quantum gravity theories.

Quantum gravity, string theory, and an Integrated Theory of Everything. Quantum gravity, string theory, and an Integrated Theory of Everything are one because they unify all known physical phenomena from the infinitely small or Planck cube scale (quantum mechanics) to the infinitely large or Super Universe scale (Einstein's General Relativity).

Quantum mechanics. Quantum mechanics is the branch of mechanics based on a quantum or a unit of energy/mass. Each of the five fundamental matter particles (up quark, down quark, electron, zino, and photino) and each of the six force particles (gravitational, electromagnetic, strong, weak, super, and Higgs) exist within a Planck cube which is the quantum or unit of matter particle, force particle, and space.

Quark-nova. A quark-nova is a gravitational collapse of a massive neutron star to a very dense quark star (matter) and the subsequent explosion which releases stupendous matter and radiation energy.

Quark star (matter). A quark star (matter) is a star denser than a neutron star and consists of five fundamental matter particles (up quark, down quark, electron, zino and photino) and their five associated Higgs forces. It contains between several and one million solar masses and is one type of stellar black holes.

Reheating. Reheating is the generation of matter and force particles following inflation in the Extended Ultimate Free Lunch theory. There is no separate reheating phase in the proposed alternate theory, An Integrated Theory of Everything.

Rest mass energy. Rest mass energy E is defined as $E = mc^2$ where m is the particle's mass and c is the velocity of light.

Resurrected life. Resurrected life refers to our precursor universe's super supermassive quark star (matter) collapse to its associated super supermassive black hole (energy), which reset disorder from maximum to minimum and resurrected life via creation of the super force or mother particles.

Second Law of Thermodynamics. The Second Law of Thermodynamics states disorder increases irreversibly with time and provides an arrow of time direction.

Singularity. A singularity is a point at the center of a single Planck cube. At the start of our universe, all our universe's matter and force particles were in the form of near infinitely compressible super force particles in a doughnut-shaped singularity.

Solar mass. The mass of our sun or one solar mass equals approximately 2×10^{30} kilograms.

Standard candle. A standard candle is an astronomical object with a consistent luminosity or brightness, for example, a type Ia supernova.

Start of universe. The start of our universe, or the start of our big bang, or the start of cosmology, occurred at time $t = 0$.

Stellar black hole (energy). A stellar black hole (energy) consists of super force energy in a doughnut-shaped singularity.

Stellar black holes. Stellar black holes are stars which are so dense not even light can escape from them. Six types of stellar black holes are quark star (matter), supermassive quark star (matter), super supermassive quark star (matter), its associated super supermassive black hole (energy), super super supermassive quark star (matter), and its associated super super supermassive black hole (energy).

Strong force. The strong force is the nuclear binding force between up quarks and down quarks within a proton or neutron and between protons and neutrons in a nucleus.

String. All five matter particles (up quark, down quark, electron, zino, and photino) and all six force particles (gravitational, electromagnetic, strong, weak, super, and Higgs) are strings at the fifth building block level. A string is visualized as a thin sticky band wrapped around a Planck cube sized beach ball with hill's and valley's amplitude displacement and frequency.

String theory. The goal of string theory is to unify all known physical phenomena from the infinitely small or Planck cube scale to the infinitely large or Super Universe scale.

Super force. The super force is the mother particle of five matter particles and five force particles. At the start of our universe, all matter and force particles were in the form of near infinitely compressible super force particles in a doughnut-shaped singularity.

Supermassive quark star (matter). A supermassive quark star (matter) consists of five fundamental matter particles (up quark, down quark, electron, zino, and photino) and their associated Higgs forces. It is a stellar black hole having an energy/mass between one million and ten billion solar masses. At the center of each of our universe's 100 billion galaxies is a supermassive quark star (matter); see table 5.

Supernova. A supernova is a stellar explosion. Two types of supernovas are Ia and II. In a type Ia, one of the binary stars is a white dwarf which accretes mass from its companion star, reaches the critical Chandrasekhar limit of 1.38 solar masses, and explodes with a standard brightness (standard candle). A type II supernova is the gravitational collapse of a massive star to a dense neutron star and the subsequent explosion which releases enormous matter and radiation energy.

Super supermassive black hole (energy). A super supermassive black hole (energy) is a stellar black hole having an energy/mass of 10^{24} solar masses, or the energy/mass which created our universe. Our precursor universe's super supermassive quark star (matter) collapsed to its associated super supermassive black hole (energy) and created the big bang of our universe.

Super supermassive quark star (matter). A super supermassive quark star (matter) consists of five fundamental matter particles (up quark, down quark, electron, zino, and photino) and their associated Higgs forces. It is a stellar black hole having an energy/mass between ten billion and 10^{24} solar masses.

Super super supermassive quark star (matter)/black hole (energy). A super super supermassive quark star (matter) and its associated super super supermassive black hole (energy) are stellar black holes having energy/masses much greater than 10^{24} solar masses. A super super supermassive quark star (matter) collapsed to its associated super super supermassive black hole (energy) and created our precursor universe; see table 5.

Super super super supermassive black hole (energy). The super super super supermassive black hole (energy) had an energy/mass of 10^{144} solar masses and created the Super Universe, see table 5.

Super Universe. Nested within the Super Universe are precursor universes, nested within precursor universes are universes, and nested within universes are galaxies. The Super Universe's volume is 10^{120} larger than our universe's volume.

Surface hill's and valley's amplitude displacement and frequency. Surface hill's and valley's amplitude displacement and frequency are the amplitude and frequency modulation of a Planck cube sized beach ball which defines the energy/mass of individual matter and force particles.

Systems engineering. Systems engineering is an interdisciplinary field that integrates complex systems such as radar, a space station, or an integrated Theory of Everything (TOE).

Theory of Everything (TOE). The TOE unifies all known physical phenomena from the infinitely small or Planck cube scale to the infinitely large or Super Universe scale.

Thought experiment. A thought experiment is an imaginary experiment to evaluate a theory.

Three springs potential energy model. The Planck cube sized beach ball's potential energy/mass can be modeled by three springs connected together at the Planck cube's center as shown in Figure 11.

Time t = 0. Our universe's start time is t = 0.

Ultimate Free Lunch. The Ultimate Free Lunch is the prevailing cosmological theory which states nothing existed before the big bang. That is, the near infinite energy/mass of our universe was created from nothing, or more precisely, from random energy fluctuations or noise before the big bang.

Underweight porcupine. A model of a matter particle and its associated Higgs force is an underweight porcupine with overgrown spines as shown in Figure 7.

Uniform distribution of matter. Matter was uniformly distributed during our universe's expansion from the end of inflation to 30,000 years. Samples any where in our universe contained the same number of matter particles.

Universe expansions. Our universe had four time sequential expansion phases: within our universe's first Planck cube, inflation, uniform distribution of matter, and nonuniform distribution of matter. Both the uniform and nonuniform distributions of matter expansion were driven by dark energy or Higgs force particles.

Universes. Each universe consists of approximately 100 billion galaxies. Within a precursor universe are parallel universes.

Up quark. The up quark is one of three fundamental atomic fourth level matter particles (up quark, down quark, and electron) and has a fractional (+ 2/3) positive charge.

Weak force. The weak force transforms, for example, down quarks into up quarks and vice versa.

White dwarf star. A white dwarf star is a dense star which is less dense than a neutron star and consists primarily of unbounded atomic nuclei and electrons.

White hole. A white hole is the reverse of a black hole. A black hole swallows matter and energy whereas a white hole expels it. Matter and energy can escape from a white hole but not from a black hole.

Wormhole. A wormhole is a singularity which connects two universes. For example, our precursor universe's super supermassive black hole (energy) doughnut-shaped singularity was connected to our universe's white hole, or big bang.

Zino. The zino is one of two fundamental dark matter particles. The second is a photino.

Further Readings

Greene, Brian. *The Elegant Universe*. New York: Vintage Books, 2000.

——. *The Fabric of the Cosmos*. New York: Vintage Books, 2005.

Guth, Alan H. *The Inflationary Universe*. New York: Perseus Publishing, 1997.

Randall, Lisa. *Warped Passages*. New York: Harper Perennial, 2006.

Rees, Martin, ed. *Universe*. New York: DK Publishing, 2005.

M050112.doc

Appendix A An Integrated Theory of Everything Research Article

Copyright © May 2012 by Antonio A. Colella

Abstract

An Integrated Theory of Everything (TOE) unifies all known physical phenomena from the infinitely small or Planck cube scale to the infinitely large or Super Universe scale. Each of 129 fundamental matter and force particles is represented by its unique string in a Planck cube. Any object in the Super Universe can be represented by a volume of contiguous Planck cubes containing fundamental matter or force particle strings. Super force string singularities at the center of Planck cubes existed at the start of the Super Universe, all precursor universes, and all universes including our universe.

The foundations of the theoretical Integrated TOE are the following twenty independent existing theories; string, particle creation, inflation, spontaneous symmetry breaking, Higgs forces/supersymmetric Higgs particles, superpartner and quark decays, neutrino oscillations, dark matter, universe expansions, dark energy, messenger particle operational mechanism, relative strengths of forces, Super Universe (multiverse), stellar black holes, black hole entropy, arrow of time, cosmological constant problem/

nested universes, black hole information paradox, baryogenesis, and quantum gravity. The premise of an Integrated TOE is without sacrificing their integrities; these twenty independent existing theories are replaced by twenty interrelated amplified theories.

An Integrated TOE was developed by a top down, iterative, systems engineering technique which selectively amplified each independent existing theory to integrate it with interrelated theories without sacrificing the existing theory's integrity. An example of requirement amplification was matter particle creation theory was amplified to be time synchronous with inflation theory. The results of an Integrated TOE were summarized in Table IV, Primary interrelationships between twenty amplified theories.

DOI: Subject Areas: Cosmology, Supersymmetric Higgs Particles, and String Theory

Introduction

An Integrated TOE unifies all known physical phenomena from the infinitely small or Planck cube scale to the infinitely large or Super Universe scale. Each matter and force particle exists within the universe's fundamental building block, the Planck cube. Any universe object is representable by a volume of contiguous Planck cubes. The Planck cube is the quantum or unit of matter particle, force particle, and space.[102] An Integrated TOE unifies 16 Standard Model particles, 16 supersymmetric particles, 32 anti-particles, their 64 associated supersymmetric Higgs particles, and the super force or mother particle for 129 particles.

The foundations of an Integrated TOE are the following twenty independent, existing theories; string, particle creation, inflation, spontaneous symmetry breaking, Higgs forces/supersymmetric Higgs particles, superpartner and quark decays, neutrino oscillations, dark matter, universe expansions, dark energy, messenger particle operational mechanism, relative strengths

Appendix A An Integrated Theory of Everything Research Article

of forces, Super Universe (multiverse), stellar black holes, black hole entropy, arrow of time, cosmological constant problem/nested universes, black hole information paradox, baryogenesis, and quantum gravity. These twenty independent existing theories were developed by physicists primarily for internal integrity with minor emphasis on interrelated theories.

An Integrated TOE was developed as follows. A top down, iterative, systems engineering technique selectively amplified each independent existing theory to integrate it with interrelated theories without sacrificing the existing theory's integrity. For example, the key Higgs forces/supersymmetric Higgs particles theory was amplified to include; Higgs force particles were residual super force particles, matter particles and their associated Higgs forces were one and inseparable and spontaneous symmetry breaking was bidirectional.

The foundations of an Integrated TOE include twenty independent existing theories and their accepted experimental data or observations. An Integrated TOE's predictions are experimentally distinguishable from existing knowledge. This article's motivation and justification is the 20 independent existing theories are now an Integrated TOE consisting of 20 interrelated amplified theories as summarized in Table IV, Primary interrelationships between twenty amplified theories.

String theory

Each of 129 fundamental matter and force particles is represented by its unique string or associated Calabi-Yau membrane in a Planck cube. A string or associated Calabi-Yau membrane's energy/mass is primarily a function of its diameter and secondarily its hill's and valley's amplitude displacement and frequency. The big bang's near zero diameter singularity of superimposed super force strings consisted of our universe's near infinite energy. Any object in the Super Universe can be represented by a volume of contiguous Planck cubes containing fundamental matter or force particle strings. Super force string singularities at the center of Planck cubes existed at the start of the Super Universe, all precursor universes, and all universes including our universe (see Cosmological constant problem/nested universes).

The inertially stabilized X_u, Y_u, Z_u universal rectangular coordinate system of Fig. 1 originates at our universe's big bang at $x_u = 0, y_u = 0, z_u = 0, t = 0$.[103] A Planck length ($l_p$ = 1.6 x 10^{-35} meters) cube is centered at x_u, y_u, z_u at time t with the cube's $X_p, Y_p,$ and Z_p axes aligned with the X_u, Y_u, Z_u axes. Any point within the Planck cube is identified by x_p, y_p, z_p coordinates measured from the cube's center with velocity components $v_{xp}, v_{yp},$ and v_{zp}. At t = 0, our universe consisted of a super force singularity centered in a Planck cube at $x_u = 0, y_u = 0,$ and $z_u = 0$. At the present time t = 13.7 billion years, our universe consists of approximately 10^{185} contiguous Planck cubes.

Each of the 129 Standard Model/supersymmetric particles listed in Table I exists within a Planck cube and is equivalently represented by a dynamic phantom point particle, its unique string, or its associated Calabi-Yau membrane. In traditional string theory descriptions, a one brane vibrating string generates a two brane Calabi-Yau membrane over time. A zero brane dynamic phantom point particle generates quantized particle positions for both a one brane vibrating string and a two brane Calabi-Yau membrane over time. String theory's six extra dimensions are the dynamic phantom point particle position (x_p, y_p, z_p) and velocity (v_{xp}, v_{yp}, v_{zp}) coordinates.

A basic Calabi-Yau membrane type is a Planck cube sized beach ball to which periodic surface hills' and valleys' are added to synthesize any particle. A string can be visualized as a thin sticky band wrapped around a Calabi-Yau membrane. For example, a circle with periodic hill's and valley's is the string associated with a beach ball membrane with periodic surface hill's and valley's.

Appendix A An Integrated Theory of Everything Research Article

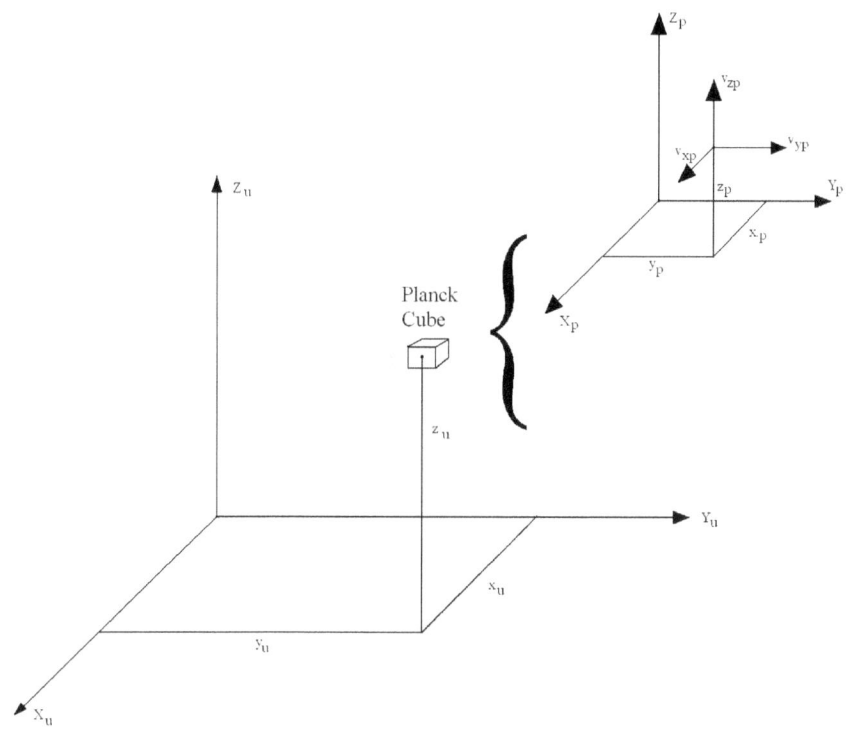

FIG. 1. Universal rectangular coordinate system.

A membrane's potential energy/mass can be represented by three springs aligned along the X_p, Y_p, and Z_p axes, and connected together at $x_p = 0$, $y_p = 0$, and $z_p = 0$. A Calabi-Yau membrane's energy/mass is primarily a function inversely proportional to its diameter and secondarily directly proportional to its surface hill's and valley's amplitude displacement and frequency.[104] A particle's energy/mass is amplified from two to three Calabi-Yau membrane or string parameters by addition of the diameter parameter. Diameter defines the basic energy/mass level whereas the amplitude displacement and frequency fine tunes it. A Calabi-Yau membrane just touching the Planck cube sides with zero amplitude displacement and frequency represents zero tension or energy/mass. A range of amplitude displacements and frequencies about this level defines the 32 fundamental matter and force particles' energy/masses, from the lightest photon (zero) to the top quark (173 GeV) to supersymmetric particles (100 to 1500 GeV).

In contrast, the big bang's near zero diameter singularity of superimposed super force strings consisted of our universe's near infinite energy of approximately 10^{54} kilograms, or $10^{24}\,M_\odot$, or 10^{90} eV.[105] The super force singularity was a rotating, charged, doughnut shaped Calabi-Yau membrane or Kerr-Newman black hole. Pauli's exclusion principle states no two matter particles can have identical quantum numbers, which was assumed equivalent to occupying the same Planck cube.[106] In contrast, Pauli's exclusion principle permits force particles to exist within the same Planck cube such as the super force singularity. This integrates string with particle creation and stellar black holes theories (see Table IV).

A proton consisting of quarks, photons, and gluons can be represented by a volume of contiguous Planck cubes (see Fig. 5). An atom can be represented by a volume of contiguous Planck cubes consisting of protons, neutrons, and orbital shell electrons. By extension, any object in the Super Universe (e.g. molecule, encyclopedia, star, galaxy, or the Super Universe) can be represented by a volume of contiguous Planck cubes containing fundamental matter or force particles. The contiguous Planck cubes can be visualized as extremely small, cubic, Lego blocks. Quantized time is represented by Planck time.

Proposed standard/supersymmetric particle symbols

Two reasons for replacing inadequate existing symbols with proposed symbols are; explicit and amplified Higgs particle representation and elimination of existing symbol ambiguities via standardization of subscripts and capitals.

Table I shows the proposed symbols with Standard Model particles on the left and supersymmetric particles on the right. The subscript xx explicitly identifies a specific matter or force particle (e.g. the number 11 identifies the up quark p_{11}). Adding sixteen to the Standard Model particle subscript identifies its supersymmetric partner (e.g. up squark p_{27}). Replacing p with h identifies the associated Higgs particle (e.g. h_{11} is the Higgs force associated with the up quark p_{11}). An anti-particle is identified by the subscript bar (e.g. the anti-up quark is p_{11bar}). The proposed symbols are different than existing symbols. For example the up quark p_{11} replaces u, the down quark p_{10} replaces d, the up squark p_{27} replaces a u with a tilde over it, the up quark anti-particle p_{11bar} replaces a u with a bar over it, and the photon p_{16} replaces γ.

The first reason for the proposed symbols is explicit Higgs particle symbols are not available in existing symbols. In the proposed symbols, there is a Higgs particle for each matter and force particles. Since there are 16 Standard Model particles, 16 supersymmetric particles, and 32 anti-particles, there are 64 supersymmetric Higgs particles. Each matter particle has an associated Higgs force and each force particle has an associated Higgsino or Higgs matter particle. Explicit Higgs particles are essential because as subsequently described, the sum of Higgs force energies associated with eight permanent matter particles is dark energy and Higgsinos experience spontaneous symmetry breaking.

The second reason for the proposed symbols is elimination of existing symbol ambiguities via standardization of subscripts and capitals. The first example is eight types of gluons p_2 are explicitly represented by; p_{2a}, p_{2b}, p_{2c}, p_{2d}, p_{2e}, p_{2f}, p_{2g}, and p_{2h}. Eight explicit gluon symbols are not available in existing symbols.

A second example is the photon p_{16} which is divided into two types; p_{16a} for electromagnetic radiation and p_{16b} for force carrier. Electromagnetic radiation is further subdivided into gamma ray p_{16a1}, X rays p_{16a2}, etc. The photon symbol γ illustrates ambiguities of existing symbols because all electromagnetic and the specific gamma ray radiation are defined by γ. In addition, a force carrier photon is not defined in existing symbols and annihilation of matter and anti-matter particles produces super force particles (p_{sf}) not electromagnetic radiation (γ).

A third example is there are seventeen different types of super force particles which condense into seventeen different matter particles. The seventeen super forces types are identified for example by p_{sfp11} where the subscripts (sf) signify super force and the following subscripts (e.g. p11) signify the condensed matter particle. There is only one super force in existing symbols.

A fourth example is total particle energy/mass is represented by upper case letter symbols, for example, total up quark energy/mass is P_{11}. The subsequently described big bang time line of Fig. 2 uses total energy/mass for 32 matter and force particles. Total energy/mass for an individual particle is not available in existing symbols.

Appendix A An Integrated Theory of Everything Research Article

TABLE I. Proposed Standard Model/supersymmetric particle symbols.

Symbol	Standard Model	Matter	Force	Symbol	Supersymmetric	Matter	Force
p_1	graviton		x	p_{17}	gravitino	x	
p_2	gluon		x	p_{18}	gluino	x	
p_3	top quark	x		p_{19}	top squark		x
p_4	bottom quark	x		p_{20}	bottom squark		x
p_5	tau	x		p_{21}	stau		x
p_6	charm quark	x		p_{22}	charm squark		x
p_7	strange quark	x		p_{23}	strange squark		x
p_8	muon	x		p_{24}	smuon		x
p_9	tau-neutrino	x		p_{25}	tau-sneutrino		x
p_{10}	down quark	x		p_{26}	down squark		x
p_{11}	up quark	x		p_{27}	up squark		x
p_{12}	electron	x		p_{28}	selectron		x
p_{13}	muon-neutrino	x		p_{29}	muon-sneutrino		x
p_{14}	electron-neutrino	x		p_{30}	electron-sneutrino		x
p_{15}[107]	W/Z bosons		x	p_{31}	wino/zinos	x	
p_{16}	photon		x	p_{32}	photino	x	

16 Standard Model $p_1 \ldots p_{16}$
16 Supersymmetric $p_{17} \ldots p_{32}$
32 anti-particles $p_{1bar} \ldots p_{32bar}$
64 Higgs particles $h_1 \ldots h_{32}, h_{1bar} \ldots h_{32bar}$
1   super force (mother) p_{sf} (17 types)
───
129 total

A fifth example is there are seventeen different super force densities which condense into seventeen different matter particles. The seventeen super force densities are identified for example by P_{sfdp11} where the subscripts (sfd) signify super force density and the following subscripts (e.g. p11) signify the condensed matter particle. Seventeen super force densities are subsequently described in the spontaneous symmetry breaking section. Only one super force density is available in existing symbols.

Particle creation/Inflation

The big bang created our universe's 128 particles from the super force having energy of 10^{54} kilograms. Matter creation was time synchronous with both the inflationary period start time and the one to seven Planck cubes energy to matter expansion. By t = 100 seconds, all super force energy had condensed into eight permanent matter particles and their eight associated Higgs force energies. Fig. 2 Big bang shows creation of our universe's 128 particles from the super force P_{sf} having energy of 10^{54} kilograms.[108] Upper case letters are exclusively used because particle creation involves total particle energy/mass, for example, total up quark energy/mass is P_{11}. Total energy/mass (e.g. P_{11}) consists of three types of energies: rest mass, kinetic (translational and rotational), and potential (gravitational, electromagnetic, nuclear binding) energies for each up quark particle p_{11} multiplied by the number of up quark particles n_{11}. Matter particles are described by energy/mass whereas force particles are described by energy. Up quark energy density P_{11d} is total up quark energy/mass P_{11} divided by our universe's volume at the time of up quark creation.

Appendix A An Integrated Theory of Everything Research Article

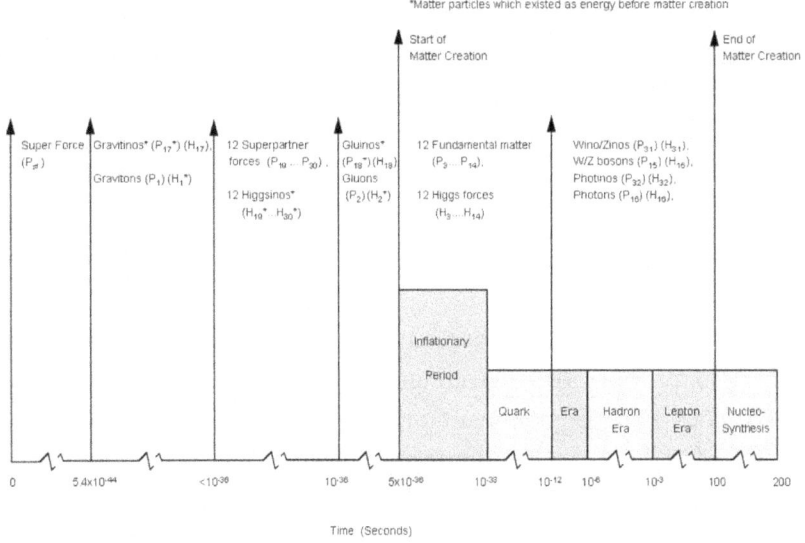

FIG. 2. Big bang.

Fig. 2 shows creation of energy/masses for gravitinos* (P_{17}*)/gravitons (P_1) at t = 5.4 x 10^{-44} seconds and gluinos* (P_{18}*) /gluons (P_2) at t = 10^{-36} seconds. The asterisk (*) signifies matter particles which existed as energy before matter creation. Twelve superpartner force energies (P_{19}....P_{30}) were created at < 10^{-36} seconds and consisted of X bosons. Grand Unified Theory (GUT) bosons included 8 gluons p_2, 3 W/Z bosons p_{15},[109] and photons p_{16}. A portion of the GUT bosons and their superpartners, gluons and gluinos, condensed at t = 10^{-36} seconds. A second portion consisting of W/Z bosons, wino/zinos, photons, and photinos condensed at t = 10^{-12} seconds.

Matter creation theory was amplified to be time synchronous with both the inflationary period start time (5 x 10^{-36} seconds) and the one to seven Planck cubes energy to matter expansion. This eliminated a separate reheating phase following inflation. Since Pauli's exclusion principle prohibited matter particles from existing within the same Planck cube, matter did not exist when our universe was smaller than one Planck cube or when our universe's radius was .8 x10^{-35} meters, see Fig. 3.[110] The one to seven Planck cubes energy to matter expansion consisted of six contiguous Planck cubes attached to the six faces of our universe's original Planck cube. The original Planck cube contained superimposed super force particles whereas the six contiguous cubes contained six newly created matter particles. Following the start of matter creation, gravitinos* (P_{17}*), gluinos* (P_{18}*), and 12 fundamental matter (6 quarks and 6 leptons) particles (P_3....P_{14}) energy/masses were condensed to matter particles. At t = 10^{-12} seconds, W/Z bosons (P_{15}), winos/zinos (P_{31}) and photino (P_{32}) energy/masses were condensed to matter particles. This integrated inflation and particle creation theories, (see Table IV).

Particle/anti-particle pairs condensed from super force energy and evaporated back to the super force. As our universe expanded and cooled this baryogenesis process was predominantly from energy to matter rather than to anti-matter (see Spontaneous symmetry breaking/Higgs forces and Baryogenesis sections). Particles/anti-particles were the intermediate or false vacuum state (quantum fluctuations) prior to the permanent matter plus true vacuum state. During matter creation (5 x 10^{-36} to 100 seconds), our universe consisted of a time varying particle soup. The end of matter creation was defined as 100 seconds because by: 10^{-3} seconds, up and down quarks formed protons and neutrons; 1 second, neutrinos decoupled from matter; 100 seconds, only electrons remained following electron anti-electron annihilations.[111] By t = 100 seconds, all super force energy had condensed into eight permanent matter particles and their eight associated Higgs force energies. Also at t = 100 seconds, nucleosynthesis began.

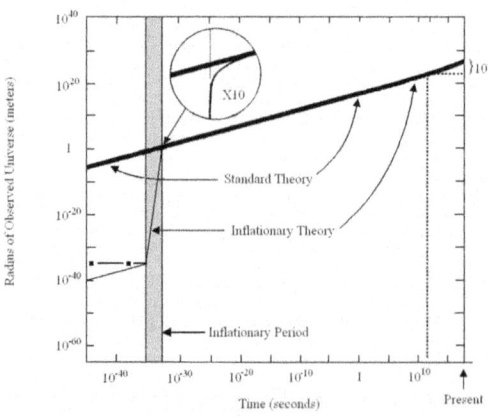

FIG. 3. Size of universe in the standard and inflationary theories.

Appendix A An Integrated Theory of Everything Research Article

Spontaneous symmetry breaking/Higgs forces

The process of generating 17 matter particles and their 17 associated Higgs forces is spontaneous symmetry breaking or the Higgs mechanism.[112] The sum of eight permanent Higgs forces' energies associated with eight permanent matter particles: atomic matter (up quark, down quark, electron); dark matter (zino, photino); and neutrino matter (tau-neutrino, muon-neutrino, electron-neutrino) constitutes dark or vacuum energy.

Amplifications of Higgs force theory are: Higgs force particles are residual super force particles containing characteristics (e.g. mass, charges, spin) of their associated matter particles; matter particles and their associated Higgs forces are one and inseparable; spontaneous symmetry breaking is bidirectional supporting condensations from and evaporations to the super force; super force condensations occur for 17 matter particles and their associated Higgs forces; and the sum of eight permanent Higgs force energies is dark energy.

Fig. 2 shows energy/masses of 32 matter and force particles designated as $P_1....P_{32}$. These included gravitons P_1, gluons P_2, twelve fundamental matter particles ($P_3....P_{14}$), W/Z bosons P_{15}, photons P_{16}, 4 supersymmetric matter particles (P_{17}^*, P_{18}^*, P_{31}, and P_{32}), and 12 supersymmetric force particles ($P_{19}....P_{30}$) energy/masses. The 17 Higgs force energies ($H_3....H_{14}$, H_{17}, H_{18}, H_{31}, H_{32}, H_{15}) were super force energy residuals following condensations of 12 fundamental matter, four supersymmetric matter, and W/Z bosons energy/masses. There were also 15 Higgs matter particles (14 Higgsinos* and 1 Higgsino) energy/masses (H_1^*, H_2^*, H_{19}^*.... H_{30}^*, H_{16}) for a total of 32 Higgs particles. Thirty two anti-particles condensed with their 32 associated Higgs particles at the same temperature and time as their identical energy/mass particles but were not explicitly shown in Fig. 2 because baryogenesis eliminated them. Thus, super force (P_{sf}) energy equaled 32 Standard Model/supersymmetric matter and force particles and their 32 associated Higgs particle energy/masses or,

$P_{sf} = (P_1 + H_1^*)....(P_{32} + H_{32})$. From Fig. 2 at $t = 5.4 \times 10^{-44}$ seconds, one super force pair's energy $(P_1 + H_1^*)$ condensed into gravitons' energy (P_1) and its associated Higgsino* energy/mass (H_1^*) and a second super force pair $(P_{17}^* + H_{17})$ condensed into the gravitinos* energy/mass (P_{17}^*) and its associated Higgs force energy (H_{17}). At $t = 10^{-36}$ seconds, a third super force pair's energy $(P_2 + H_2^*)$ condensed into gluons' energy (P_2) and its associated Higgsino* energy/mass (H_2^*) and a fourth super force pair $(P_{18}^* + H_{18})$ condensed into gluinos* energy/mass (P_{18}^*) and its associated Higgs force energy (H_{18}). At $t < 10^{-36}$ seconds, twelve super force energy pairs $[(P_{19} + H_{19}^*)....(P_{30} + H_{30}^*)]$ were created as X bosons and their associated Higgsinos.[113] During our universe's matter creation period (5×10^{-36} to 100 seconds), four supersymmetric matter energy/masses $(P_{17}^*, P_{18}^*, P_{31}, P_{32})$, their associated Higgs force energies $(H_{17}, H_{18}, H_{31}, H_{32})$, 12 fundamental matter energy/masses $(P_3....P_{14})$ and their associated Higgs force energies $(H_3....H_{14})$ created four supersymmetric matter, 12 fundamental matter, and 16 associated Higgs force particles. At $t=10^{-12}$ seconds, two super force pairs of energy $(P_{15} + H_{15})$ and $(P_{16} + H_{16})$ condensed into W/Z bosons (P_{15}), photons (P_{16}), and their two associated Higgs particles (H_{15}, H_{16}).

The up quark spontaneous symmetry breaking function is shown in Fig. 4.[114] The Z axis represents energy density of the super force (i.e. P_{sfdp11}) available for condensation to up quarks and their associated Higgs force particles. The X axis represents one Higgs force particle's energy h_{11} associated with an up quark particle p_{11}. Similarly, the Y axis represents one Higgs force particle's energy h_{11bar} associated with the anti-up quark particle p_{11bar}. Because of the early universe's baryogenesis, anti-particles quickly disappear and Fig. 4 compresses to the two dimensional Z versus X diagram shown in the Fig. 4 inset.[115] The Z axis represents: prior to condensation, the up quark and its associated Higgs force energy densities $[P_{sfdp11} = (P_{11d} + H_{11d})]$; or following condensation, the associated Higgs force energy density (H_{11d}). At the peak position, all the energy density is super force density (P_{sfdp11}). At the position shown by the ball ($h_{11} = -2$, $h_{11bar} = 0$, $Z =$

Appendix A An Integrated Theory of Everything Research Article

1.5), condensation of up quark p_{11} particles is complete, the residual energy density is H_{11d}, and the associated h_{11} is non-zero. Following condensation, the h_{11} non-zero value (-2) remains constant, (i.e. ball slowly over 13.7 billion years moves vertically down and approaches the vacuum circle for up quarks) while the associated Higgs energy density (H_{11d}) decreases as our universe expands. Each Higgs force h_{11} contains the characteristics (e.g. mass, charges, and spin) of its associated particle p_{11} and itself (see Fig. 5 Inset). The Higgs force h_{11} is visualized as a three dimensional field surrounding and inseparable from the p_{11} particle or symbolically as a single Planck cube attached to its p_{11} particle.[116]

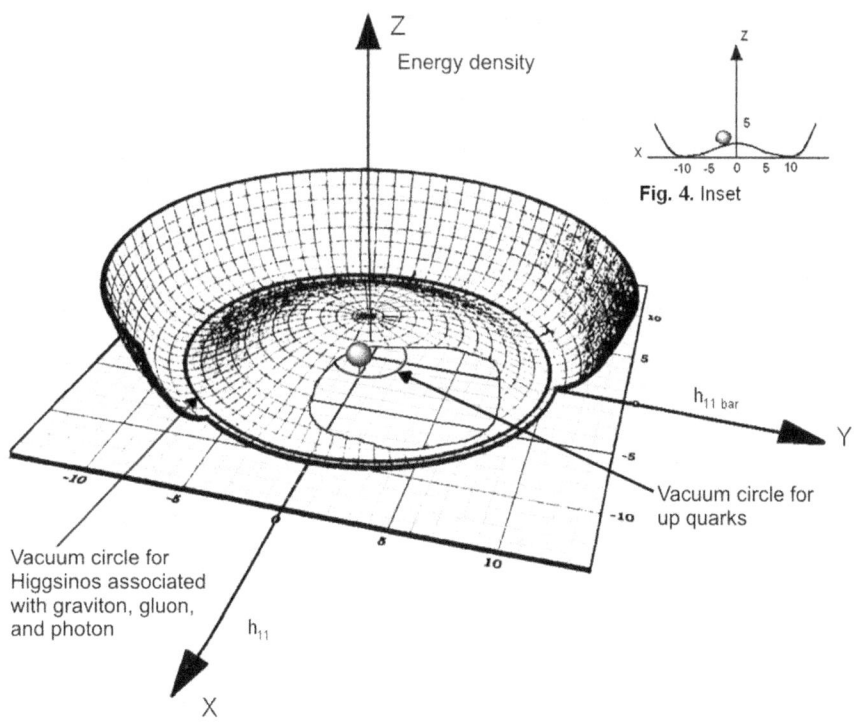

FIG. 4. Up quark spontaneous symmetry breaking function.

Super force density condensations occurred for 17 matter particles (p_3....p_{14}, p_{15}, p_{17}, p_{18}, p_{31}, p_{32}) and produced 17 associated Higgs force particles (h_3....h_{14}, h_{15}, h_{17}, h_{18}, h_{31}, h_{32}). The assumed heaviest matter particle's (e.g. gravitino p_{17}) spontaneous symmetry breaking function occurred first during matter creation. There were 17 unique spontaneous symmetry breaking functions having the generic shape of Fig. 4, which occurred at different times or temperatures during matter creation.

The false vacuum was an intermediate state where the super force condensed either to transient matter particles or particles/antiparticles and bidirectionally evaporated back to the super force. During matter creation, nine transient matter particles (top quark p_3, bottom quark p_4, charm quark p_6, strange quark p_7, tau p_5, muon p_8, gravitino p_{17}, gluino p_{18}, and W/Z bosons p_{15}) and their nine associated Higgs forces condensed from and evaporated back to the super force.[117]

The true or permanent vacuum state was space without matter or the lowest energy/temperature density state. The sum of eight permanent Higgs force energies (H_{11}, H_{10}, H_{12}, H_{31}, H_{32}, H_9, H_{13}, H_{14}) associated with eight permanent matter particles: atomic matter (up quark p_{11}, down quark p_{10}, electron p_{12}); dark matter (zino p_{31}, photino p_{32}); and neutrino matter (tau-neutrino p_9, muon-neutrino p_{13}, electron-neutrino p_{14}) constituted dark or vacuum energy. This integrates spontaneous symmetry breaking, Higgs forces/supersymmetric Higgs particles, particle creation, inflation, dark matter, dark energy, and baryogenesis theories, (see Table IV).

Each of the 129 particles was assumed to exist within a Planck cube although each may exist in a larger augmented Planck cube defined by (1_{ap}). Scattering experiments reveal quarks and leptons to be smaller than 10^{-18} meters.[118] If higher resolution scattering reveals matter particles are larger than a Planck cube, all analyses remains valid by replacing a Planck cube with an augmented Planck cube.[119]

Appendix A An Integrated Theory of Everything Research Article

Supersymmetric Higgs particles

The 32 standard and supersymmetric matter and force particles and their 32 anti-particles are supersymmetric with 64 associated Higgs particles and the latter are supersymmetric with themselves. There are three types of spontaneous symmetry breaking functions for three types of matter particles: 17 standard and supersymmetric matter particles, 3 Standard Model Higgsinos, and 12 supersymmetric Higgsinos. Higgs forces/supersymmetric Higgs particles theory is amplified from just the first type of spontaneous symmetry breaking function to all three types.

If a standard or supersymmetric particle is a matter particle (e.g. an up quark p_{11}), its associated Higgs particle is a force particle (e.g. h_{11}). If a standard or supersymmetic particle is a force particle (e.g. a graviton p_1), its associated Higgs particle is a Higgsino (e.g. h_1). If a Higgs particle is a Higgsino (e.g. the Higgsino h_1 associated with the graviton p_1), the Higgs superpartner is a Higgs force (e.g. the Higgs force h_{17} associated with the gravitino p_{17}). If a Higgs particle is a force particle (e.g. the Higgs force h_{11} associated with the up quark p_{11}), the Higgs superpartner is a Higgsino (e.g. the Higgsino h_{27} associated with the up squark p_{27}). The 32 Higgs particles associated with 32 standard and supersymmetric anti-particles are ignored because baryogenesis eliminated them.

Type 1 matter particles or the 17 standard and supersymmetric matter particles include the: top quark p_3, bottom quark p_4, charm quark p_6, strange quark p_7, down quark p_{10}, up quark p_{11}, tau p_5, muon p_8, electron p_{12}, tau-neutrino p_9, muon-neutrino p_{13}, electron-neutrino p_{14}, gravitino p_{17}, gluino p_{18}, wino/zinos p_{31}, photino p_{32}, and W/Z bosons p_{15}. These 17 standard and supersymmetric matter particles and their 17 associated Higgs forces experience spontaneous symmetry breaking as described in the previous section.

Type 2 matter particles or 3 Standard Model Higgsinos (h_1, h_2, and h_{16}) associated with three Standard Model force particles (graviton p_1, gluon

p_2, and photon p_{16}), experience spontaneous symmetry breaking as follows. The ball in Fig. 4 starts at the peak position and comes down the spontaneous symmetry breaking function along the X axis until it reaches the point where the Mexican hat intersects the XY plane (X = -10, Y = 0, Z = 0). This is on the vacuum circle for Higgsinos associated with the zero energy graviton, gluon, and photon. In effect, all a super force particle's energy is condensed to a Higgsino and none to the associated force particle (graviton p_1, gluon p_2, or photon p_{16}).

Type 3 matter particles or 12 supersymmetric Higgsinos (h_{19}....h_{30}) associated with 12 squarks and sleptons (p_{19}....p_{30}) experience spontaneous symmetry breaking as follows.[120] The ball in Fig. 4 starts at the peak position and comes down the spontaneous symmetry breaking function along the X axis to an undefined point between the maximum (X = 0, Y = 0, Z = 2) and minimum (X = -10, Y = 0, Z = 0) values. That is, a super force particle condenses into a supersymmetric Higgsino and an associated squark or slepton. The 12 squarks and sleptons are X bosons. X bosons are the latent energy (inflatons) which expanded our universe during the inflationary period and then disappeared. X bosons are to the inflation period as Higgs forces (dark energy) are to our universe's expansion following inflation. X bosons are also the intermediate force particles (W/Z_{ss} bosons) for supersymmetric (ss) particles as described in the next section. This integrates Higgs forces/supersymmetric Higgs particles with the universe expansions theory, (see Table IV).

Superpartner and quark decays/Neutrino oscillations

Intermediate force particles are W/Z bosons for Standard Model particles and supersymmetric W/Z_{ss} bosons for supersymmetric particles. Decays are a series of evaporations of matter particles and their associated Higgs forces to the super force and condensations from the super force to less massive matter particles and their associated Higgs forces. The neutral heavy lepton is a constituent of dark matter. The theory of Superpartner and quark decays is amplified to include supersymmetric W/Z_{ss} bosons and simultaneous decay of matter particles with their associated Higgs forces.

Appendix A An Integrated Theory of Everything Research Article

The heaviest matter particles condensed directly from the super force. Lighter matter particles were created primarily via a heavier particle's decay. Decays were mediated by gauge interactions. Heavier matter particles decayed in a cascading process to lower energy/mass matter particles and intermediate force particles. Intermediate force particles were W/Z bosons for Standard Model particles and supersymmetric W/Z$_{ss}$ bosons (X bosons) for supersymmetric particles. For example, in a Beta minus decay, the W$^-$ boson decays to an electron and an anti-electron-neutrino. Similarly, the supersymmetric W/Z$_{ss}$ boson decays to a quark and lepton. Superpartners decayed into lower energy/mass superpartners. The decay chain ended with the stable Lightest Supersymmetric Particle (LSP)[121] and Standard Model particles.[122]

Heavy quarks decayed into lower energy/mass quarks and W bosons defined by the Cabibbo-Kobayashi-Maskawa (CKM) matrix. Quark decays were described by amplified weak force Feynman diagrams which consisted of evaporations of matter particles and their associated Higgs forces to the super force and condensations from the super force to less massive matter particles and their associated Higgs forces. Amplified Beta minus decay was as follows. The down quark p_{10} and its associated Higgs force h_{10} evaporated to a super force particle p_{sfp10} having energy ($p_{10} + h_{10}$). Division of energy not matter occurred as one energy portion condensed into the up quark p_{11} and its associated Higgs force h_{11}, and a second portion condensed into the W$^-$ particle p_{15} and its associated Higgs force h_{15}. Within 10^{-25} seconds, the W$^-$ and its associated Higgs force evaporated back to a super force particle having energy ($p_{15} + h_{15}$). This energy then condensed into an electron p_{12}, its associated Higgs force h_{12}, an anti-electron-neutrino p_{14bar}, and its associated Higgs force h_{14bar}. This integrates superpartner and quark decays and spontaneous symmetry breaking theories, (see Table IV)

There were three neutrino flavors: electron-neutrino, muon-neutrino, and tau-neutrino. Neutrinos oscillated between three flavors via the seesaw model using a neutral heavy lepton (NHL). According to this seesaw model, neutrino mass was $(m_D)^2/M_{NHL}$, where m_D was the Standard Model Dirac mass (i.e. p_{14}, p_{13}, p_9) and M_{NHL} was the neutral heavy lepton mass.[123] The neutral heavy lepton appeared in some Standard Model extensions and was assumed to be the stable fourth family neutrino and a constituent of dark matter.[124] This integrates neutrino oscillations, spontaneous symmetry breaking, and dark matter theories, (see Table IV).

Dark matter

Dark matter consisted of zinos and photinos. Dark matter agglomeration formed the framework of galaxies.

Superpartners decay into the LSP and Standard Model quarks and leptons. A prime candidate for dark matter is the LSP or neutralino which is an amalgam of the zino p_{31}, photino p_{32}, and possibly other particles including Higgsinos.[125] Dark matter is assumed to consist of two supersymmetric matter particles (p_{31}, p_{32}) and neutral heavy leptons (either p_{31} or p_{32}).

Start of dark matter agglomeration defined the transition between our universe's uniform and non-uniform distribution of matter expansions. Following this transition, galactic regions were represented by static spatial cubes whereas intergalactic regions were represented by dynamic spatial cubes. Assuming a dark matter agglomeration start time of 30,000 years,[126] the Fig. 3 dotted lines show a 10^4 universe range factor expansion from 30,000 years (~ 10^{12} seconds) to the present.

Dark matter agglomeration formed the framework of galaxies. Between 30,000 and 380,000 years dark matter clumped together, whereas electrically charged matter particles did not. At 380,000 years, electrically neutral atoms formed and began clumping around the dark matter framework.[127]

Appendix A An Integrated Theory of Everything Research Article

Universe expansions

There were four sequential universe expansions. Entropy increase of the super force and its derivatives drove the expansion within our universe's first Planck cube. X bosons' (inflatons) latent heat drove the inflationary period's exponential expansion. Dark energy drove both the uniform and non-uniform distribution of matter expansions. String theory's seventh extra dimension was the product of our universe's non-uniform distribution of matter expansion rate and the graviton's intergalactic propagation time. Universe expansions theory was amplified to include expansion within our universe's first Planck cube and identification of X bosons (12 squarks and sleptons) as the latent heat source during inflation.

During the first expansion, our universe's size expanded from a doughnut shaped singularity at $t = 0$, to a sphere with a radius of $.8 \times 10^{-35}$ meters at the start of matter creation (Figs. 2 and 3). Entropy increase of the super, gravitinos*, gravitons, 12 superpartner forces, gluinos*, gluons, and 16 associated Higgs particles drove this expansion similar to the loosening of a smaller than a Planck cube sized knot of vibrating strings.

The inflationary period expansion was similar to a water container freezing and bursting. More energy exists in liquid than frozen water. When water freezes, its temperature remains constant and latent heat is released. X bosons (12 squarks and sleptons or 12 superpartner forces) were the latent heat energy source during inflation.[128] During the one to seven Planck cube expansion, six matter particles were created (i.e. condensed or froze) and expelled from the original Planck cube to the surrounding Planck cube shell. Then, the first matter shell was pushed out to enable creation of the second matter Planck cube shell. This process continued until the end of inflation when enough Planck cubes existed for all matter particles.

Universe expansion occurred from 10^{-33} seconds to 30,000 years for the uniform distribution of matter and from 30,000 years to the present for the non-uniform distribution of matter. Dark energy (i.e. Higgs forces) drove both the uniform and non-uniform distribution of matter expansions.

Our universe's non-uniform distribution of matter expansion can be represented by a marbles/dough/balloon model consisting of marbles mixed in electromagnetically transparent rising dough in a balloon. Space between galaxies expands whereas space within galaxies does not. The rigid marbles (galaxies) do not expand, whereas the dough (intergalactic space) and the balloon (our universe) expand.[129]

Einstein's general relativity representation of static galactic spatial squares (cubes) on a rubber fabric must transition into dynamic spatial squares of intergalactic regions. Newton's gravitational force equation ($F=Gm_1m_2/r^2$) is valid for galactic regions. For intergalactic regions the radius (r) must be amplified as follows. The radius (r) consists of two components $r_1 + e_r t_i$. The first constant component (r_1) is the initial radius between two masses in two galaxies at a graviton's emission time. The second variable component ($e_r t_i$) is our universe's non-uniform distribution of matter expansion rate (e_r)[130] multiplied by the graviton's intergalactic propagation time (t_i). This product is string theory's seventh extra dimension which dilutes the intergalactic gravitational force because of our universe's non-uniform distribution of matter expansion. This integrates universe expansions with particle creation, inflation, and Higgs forces/supersymmetric Higgs particles theories (see Table IV).

Dark energy

By the end of matter creation or t = 100 seconds, our universe consisted of baryonic matter (4.6%), cold dark matter (22.8%), and dark energy (72.6%), and these percentages remained constant for 13.7 billion years. The cosmological constant was proportional to vacuum or dark energy density. Dark energy density was the sum of eight permanent Higgs force densities.

By t = 100 seconds, only eight permanent matter particles and their Higgs forces (dark energy) remained. Following t = 100 seconds, baryonic matter could be changed only by big bang, stellar, or supernova nucleosynthesis which transformed neutrons into protons and vice versa. Nucleosynthesis changed total up and down quark rest mass without significantly changing total baryonic energy/mass. This was because only 1% percent of a proton/neutron's energy/mass was rest mass and 99% was nuclear binding energy. Also, nuclear binding energy was a fraction of total kinetic and potential energy.[131] Dark matter could not change following t = 100 seconds because of insufficient temperatures. Thus by the end of matter creation, our universe consisted of baryonic matter (4.6 %), cold dark matter (22.8%), and dark energy (72.6%), and these percentages remained constant for 13.7 billion years.

At t = 100 seconds, our universe consisted of uniformly distributed matter particles (e.g. electrons, protons, neutrons, neutrinos, and dark matter) and their Higgs forces in the space between matter particles (true vacuum). Our universe's uniform 10^{10} K temperature caused radiation emission/absorption between electrons and protons. At 380,000 years, radiation ended and neutral atoms clumped around the dark matter framework. Galaxies formed after 200 million years and the temperature of intergalactic space decreased relative to galaxies. Currently, that vacuum temperature is 2.73 K. Dark energy was a constant for 13.7 billion years, however as our universe expanded, dark energy density decreased.

The cosmological constant lambda (Λ) was proportional to the vacuum or dark energy density (ρ_Λ), or $\Lambda = (8\pi G/3c^2) \rho_\Lambda$, where G is the gravitational constant and c is the speed of light.[132] Dark energy density: was uniformly distributed in our universe; was the sum of eight permanent Higgs force densities, or $\rho_\Lambda = H_{11d}, H_{10d}, H_{12d}, H_{31d}, H_{32d}, H_{9d}, H_{13d}, H_{14d}$; and decreased with time along with the cosmological constant as our universe expanded.

Messenger particle operational mechanism

Messenger particles were amplified with embedded clock/computers as their operational mechanisms.

Particles are insufficient to constitute matter, glues are also required. Strong force glue (gluon) is required for nuclei. Electromagnetic force glue (photon) is required for atoms/molecules. Gravitational force glue (graviton) is required for multi-mass systems.[133]

Gravitational/electromagnetic

The graviton/photon clock/computer calculates Newton's gravitational or Coulomb's force and provides it to the receiving particle.

Newton's gravitational force ($F = Gm_1m_2/r^2$) and Coulomb's force ($F = Cq_1q_2/r^2$) equations have the same form, where m_1 and m_2 are two masses, q_1 and q_2 are two charges, r is the range between masses/charges, G is the gravitational constant, and C is Coulomb's constant. The graviton extracts mass m_1 and the photon extracts charge q_1 from the attached Higgs force particle (e.g. h_{11} of Fig. 5 Inset) associated with the transmitting particle (e.g. p_{11}). The Higgs force particle includes mass, charges, and spin of both the particle p_{11} and its associated Higgs force h_{11}, and messenger particle p_1, p_2, p_{15}, p_{16} templates.[134] The graviton or photon also extracts G or C in the graviton p_1 and photon p_{16} templates. The clock initiates at transmission time t_t and stops at reception time t_r. The computer calculates the range factor ($1/r^2$) as $1/[(t_r - t_t)(c)]^2$. Upon graviton/photon reception the receiving mass m_2 or charge q_2 are extracted from the Higgs force particle associated with the receiving particle. The graviton/photon clock/computer calculates Newton's gravitational or Coulomb's force and provides it to the receiving particle. This integrates messenger particles and Higgs forces/supersymmetric Higgs particles theories, (see Table IV).

Appendix A An Integrated Theory of Everything Research Article

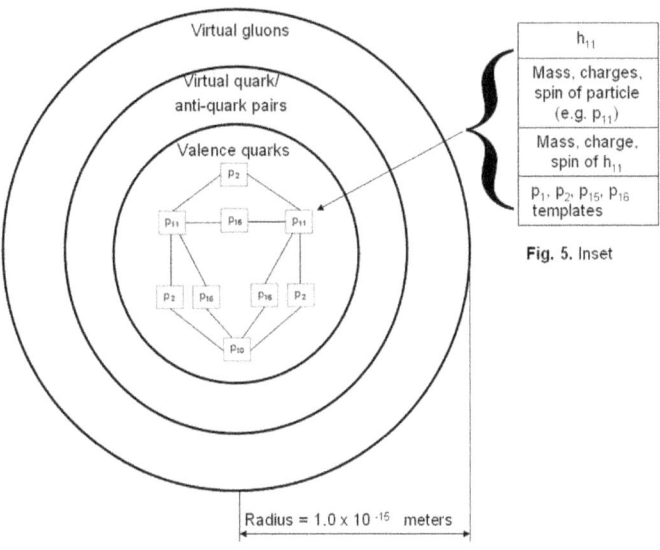

FIG. 5. Hydrogen nucleus (proton).

Strong

The gluon clock/computer calculates the strong force and provides it to the receiving quark.

The Fig. 5 hydrogen nucleus (proton) consists of contiguous Planck cubes in three nested spheres where the third sphere's radius is 1.0×10^{-15} meters. Fig. 5 is shown in two instead of three dimensions and not to scale. Up quarks p_{11}, down quarks p_{10}, photons p_{16}, and gluons p_2 exist within Planck squares. Gravitons are not included because the gravitational force is negligible within the proton radius. The proton's inner sphere contains two up and one down valence quarks. Quarks have color charges transmitted via gluons. Together, the three valence quarks are colorless. The second spherical volume contains a cloud of virtual quark/anti-quark pairs. A virtual gluon cloud exists in the third spherical volume and the two clouds adopt color charges of the valence quarks.

Quantum Chromodynamics (QCD) is strong force theory and has two major properties, confinement where the force between quarks does not diminish with separation and asymptotic freedom where the force approaches zero at short separations and quarks are free particles. Potential energy between two quarks is $V = -\alpha_s/r + kr$ and force is $F = -dV/dr = \alpha_s/r^2 - k$ where r is quark separation, k is a constant, and α_s is the running or nonlinear coupling constant which decreases with separation. The force equation has two components, a Coulomb like force (α_s/r^2) and a constant force (-k). As two confined quarks separate, the gluon fields form narrow tubes of color charge, which attract the quarks as if confined by an elastic bag. For quark separations comparable to the proton's radius, the gluon clock/computer provides the constant –k force to the receiving quark. For short quark separations less than a proton radius, the gluon clock/computer calculates the strong force using either the Coulomb term or a force versus range table lookup and provides it to the receiving quark.[135]

Relative strengths of forces/Hierarchy problem

The relative strengths of gravitational and electromagnetic/weak forces are due to propagation factor dilution ($1/r^2$) or $1/(ct)^2$ between gravitational force activation and electromagnetic/weak force creation/activation.

Column two of Table II shows relative strengths of forces. At unification, all force strengths were equal. From Fig. 2, the graviton was created at 5.4×10^{-44} seconds but activated during quark creation at approximately the beginning of the quark era or 10^{-33} seconds. At electromagnetic/weak force creation/activation time or 10^{-12} seconds, the gravitational force had already been diluted by $(t_1/t_2)^2 = (10^{-33}/10^{-12})^2$ or 10^{-42} which is the hierarchy factor. The Fig. 2 derived values in column 3 were comparable to column 2 values, considering the uncertainties of the column 2 reference and Fig. 2. This integrates relative strengths of forces with particle creation and universe expansion theories (see Table IV).

TABLE II. Relative strengths of forces.

Force	Physics handbook[136]	Figure 2 derived
Strong	1	1
Electromagnetic/weak	10^{-3} to 10^{-2}	10^{-2}
Gravitational	10^{-42}	10^{-44}

Appendix A An Integrated Theory of Everything Research Article

Super Universe

Universal laws of physics and structure were assumed across the Super Universe (multiverse). Our universe was nested in our precursor universe which was nested in the Super Universe. The Super Universe obeyed conservation of energy/mass, contained 129 particles, and had a constant dark energy to total energy/mass percentage (72.6%) just like our universe.

Stellar black holes

A stellar black hole was a quark star (matter) or black hole (energy) both of which were "black." Six types of stellar black holes were: supermassive quark star (matter), quark star (matter), super supermassive quark star (matter), its associated super supermassive black hole (energy), super super supermassive quark star (matter), and its associated super super supermassive black hole (energy). Our precursor universe's super supermassive quark star (matter)/black hole (energy) created our universe's "big bang" (white hole) via conservation of energy/mass.

Currently, a stellar black hole is defined as a region of space-time where gravity is so strong not even light can escape and having no support level below neutron degeneracy. The black hole space-time region is a three dimensional sphere which appears as a two dimensional hole. Because of black hole definition inconsistencies (e.g. a singularity is inconsistent with significant area or volume); stellar black hole theory was amplified to define a stellar black hole as a quark star (matter) or black hole (energy) both of which are "black." Their differences are a quark star (matter) has mass, volume, near zero temperature, permanence, and maximum entropy. A black hole (energy) has energy, a Planck cube singularity with minimal volume, near infinite temperature, transientness, and minimal entropy. Six types of stellar black holes are: supermassive quark star (matter), quark star (matter), super supermassive quark star (matter), its associated super supermassive black hole (energy), super super supermassive quark star (matter), and its associated super super supermassive black hole (energy).[137]

Stellar gravitational collapse occurs when internal energy is insufficient to resist the star's own gravity and is stopped by Pauli's exclusion principle degeneracy pressure. If the star's mass is less than 8 solar masses,[138] it stops contracting and becomes a white dwarf supported by electron degeneracy pressure. If the star is between 8 and 20 solar masses, it gravitationally collapses to a neutron star supported by neutron degeneracy pressure, followed by a supernova explosion. Between 20 and 100 solar masses, the star gravitationally collapses to a quark star (matter) supported by quark degeneracy pressure, followed by a quark-nova explosion.[139]

Supermassive quark stars (matter) contain 10^6 to 10^{10} solar masses. They may be "fossil quasars",[140] and their masses are proportional to their host galaxies' masses.[141] Population III stars containing hydrogen, helium, and lithium first formed approximately 200 million years after the start of our universe. These first generation stars contained up to 100 times more gas than the sun, had short lives, created over 100 billion neutron and quark stars (matter) and their protogalaxies or supernova and quark-nova remnants, and reionized our universe.[142] Over the next 13.5 billion years, by accretion of stars/matter and merger with galaxies, approximately 100 billion supermassive quark stars (matter) and their 100 billion galaxies formed in our current universe. That is, over the last 13.5 billion years, approximately 10^6 to 10^{10} solar masses fell into the original neutron and quark stars (matter).[143]

Quark stars (matter) contain between several and 10^6 solar masses. For example, quark stars (matter) having several solar masses were initially created by first generation star collapses. Their sizes were augmented by accretion of stars/matter and merger with neutron star or quark star (matter) galaxies during the next 13.5 billion years.

Super supermassive quark stars (matter) contain 10^{10} to 10^{24} solar masses. Our precursor universe's super supermassive quark star (matter)/black hole (energy) consisted of a quark star (matter) or a cold quark-gluon plasma,[144] which collapsed to its associated black hole (energy). The super supermassive quark star (matter) increased in size via accretion of stars/matter and merger with galaxies. At the 10^{24} solar mass threshold, quark degeneracy pressure was insufficient to stop further collapse. The super supermassive quark star (matter) instantaneously evaporated, deflated, and collapsed to its associated super supermassive black hole (energy). Our precursor universe's super supermassive quark star (matter)/black hole (energy) created our universe's "big bang" (white hole) via conservation of energy/mass. Super supermassive quark stars (matter)/black holes (energy) having approximately 10^{24} solar masses were to universes as supermassive quark stars (matter) were to galaxies.

Our Super Universe's super super supermassive quark stars (matter) collapsed to their associated super super supermassive black holes (energy) to create precursor universes. Super super supermassive quark stars (matter)/black holes (energy) were to precursor universes as super supermassive quark stars (matter)/black holes (energy) were to universes.

Black hole entropy

The proposed entropy formula for a quark star (matter) was proportional to the quark star's volume (r^3) and inversely proportional to a Planck cube's volume $(l_p)^3$.

Entropy of a black hole is currently defined as $S_{BH} = \eta A/(l_p)^2$ where η is a constant, A is the event horizon area, and l_p is the Planck length.[145] BH stands for either "black hole" or "Bekenstein-Hawking."

A second proposed entropy formula uses Boltzmann's equation $S = k \log \Omega$, where k is Boltzmann's constant, and Ω is the total number of different ways matter particles can arrange themselves. For the quark star (matter), the total number of ways of distributing N matter particles each in a Planck cube with volume $(l_p)^3$ within a quark star of volume $V = (4\pi r^3/3)$ is:[146]

$S = k \log \Omega$ where

$\Omega = (1/N!)(V/(l_p)^3)^N$ or

$\Omega = (1/N!)(4\pi r^3/3(l_p)^3)^N$

Arrow of time

In our universe and our precursor universe, entropy increased with time. Our universe was created by a doughnut shaped super force singularity of a super supermassive black hole (energy), surrounded by a spherical "perfect" vacuum. Our precursor universe's maximum entropy super supermassive quark star (matter) evaporated, deflated, and collapsed to the minimum entropy black hole (energy), "resurrecting" life.

In an isolated system such as our universe, the Second Law of Thermodynamics states entropy increases irreversibly with time and provides a thermodynamic arrow of time. In contrast, Einstein's Theory of General Relativity is time symmetric and apparently contradicts the Second Law of Thermodynamics. Schwarzschild's solution of Einstein's equations consists of a black hole, a white hole, and an Einstein-Rosen bridge, (i.e. wormhole or singularity) connecting the two universes.[147]

During a specific time interval within a subset volume of our universe, entropy decreased without negating our universe's Second Law of Thermodynamics.[148] A nebula's hydrogen/helium gas, dust, and plasma began ordering itself at our solar system's creation 4.6 billion years ago. Entropy decreased because life was created. Life is synonymous with low entropy and death with high entropy. Since our solar system was one of approximately 100 billion Milky Way galaxy stars and our galaxy was one of approximately 100 billion galaxies in our universe, our solar system's entropy decrease did not negate our universe's entropy increase. Similarly, entropy increased in our precursor universe whereas entropy decreased in our precursor universe's subset volume containing the super supermassive black hole (energy).

The Hawking temperature of a quark star (matter) with mass M was $T=10^{-7}$ ($M_\odot$/M) K and its life time t was approximately 10^{66} $(M/M_\odot)^3$ years, where $M_\odot$ was solar mass, and K was degrees Kelvin.[149] The larger the quark star's mass, the lower was its temperature and longer its life time. As our precursor universe's super supermassive quark star (matter) accumulated matter, its mass and life time approached near infinite whereas its temperature approached zero. Entropy increased proportional to the event horizon area in the Bekenstein-Hawking formula or to quark star volume in Boltzmann's equation. During the super supermassive quark star (matter) to black hole (energy) collapse; mass, life time, temperature, and entropy values flipped. Mass, life time, and entropy approached zero whereas temperature approached near infinite. However, total energy/mass was conserved. In the super supermassive quark star (matter), energy/mass was spread over a near infinite number of Planck cubes. In the super supermassive black hole (energy), energy was concentrated in a doughnut shaped singularity in a Planck cube. During the deflationary period collapse, each matter particle and its associated Higgs force evaporated to super force energy leaving a "perfect" vacuum in its wake. A "perfect" vacuum is completely empty whereas a true vacuum contains dark energy or Higgs forces. Since the super supermassive black hole's (energy) near infinite temperature was much higher than the surrounding "perfect" vacuum's temperature of $0°$ K, it transitioned to the white hole and initiated our universe's thermodynamic arrow of time. Our universe was created by a 10^{54} kilogram (10^{24} $M_\odot$) doughnut shaped super force singularity surrounded by a spherical "perfect" vacuum. This complied with Einstein's time symmetric Theory of General Relativity.

Fig. 6 shows our precursor universe's super supermassive quark star/black hole to our universe's big bang (white hole) transition. Fig. 6 shows time symmetry between -10^{-33} and 10^{-33} seconds. The number of super force particles was a maximum between -5×10^{-36} and 5×10^{-36} seconds. The number of super force particles decreased during inflation and reached zero at 100 seconds.[150]

Matter evaporation between < -2×10^{-33} and -5×10^{-36} seconds was the counterpart of matter creation between 5×10^{-36} and 100 seconds. Deflation occurred during all of matter evaporation whereas inflation occurred only at the beginning of matter creation. Deflation differed from inflation because its duration was longer and had two phases. The second deflation phase (-10^{-33} to -5×10^{-36} seconds) was the time reverse of inflation (5×10^{-36} to 10^{-33} seconds). That is, at -10^{-33} seconds, the super supermassive quark star (matter) consisted of a hot quark-gluon plasma with a radius of 8 meters identical to our universe at 10^{-33} seconds.[151] At -5×10^{-36} seconds, the super supermassive black hole (energy) was identical to our universe's white hole (energy) at 5×10^{-36} seconds. However, the first deflation phase was unique. The start of matter evaporation coincided with the first deflation phase at t < -2×10^{-33} seconds. Deflation of the near zero temperature super supermassive quark star (matter) began when its energy/mass reached the threshold of 10^{54} kilograms. A single electron-neutrino at the center of the super supermassive quark star (matter) was subjected to the highest pressure or temperature. This electron-neutrino and its associated Higgs force evaporated to the super force, incrementally raising the temperature of the super supermassive quark star (matter) center. This began a chain reaction which instantaneously evaporated, deflated, and collapsed the super supermassive quark star (matter) at near zero temperature first to a compact hot quark-gluon plasma at -10^{-33} seconds and then to a super supermassive black hole (energy) at -5×10^{-36} seconds. The deflationary period time was longer than the inflationary period time because it consisted of two phases instead of one.

The maximum entropy super supermassive quark star (matter) evaporated, deflated, and collapsed to the minimum entropy black hole (energy). In essence, the super supermassive black hole (energy) "resurrected" life via creation of "mother" super force particles in a subset volume of our precursor universe. Thus, the super supermassive quark star (matter)/black hole (energy) had a dual nature; destruction of structure (information) in the quark star (matter) state and resurrection of life in the black hole (energy) state.[152] This integrated the arrow of time theory with stellar black holes and black hole entropy theories (see Table IV).

Appendix A An Integrated Theory of Everything Research Article

Super Supermassive Quark Star (Matter)/Black Hole (Energy) Justification

An Integrated TOE satisfied Conservation of Energy/Mass, Einstein's Theory of General Relativity, and the Second Law of Thermodynamics for our precursor universe's super supermassive quark star (matter)/black hole (energy) transition to the big bang.

Table III compares the Ultimate Free Lunch versus an Integrated TOE. Three laws of physics are listed in column one, the Ultimate Free Lunch Theory[153] in column two, and an Integrated TOE in column three. The prevailing cosmological theory or the Ultimate Free Lunch stated nothing existed before the big bang. The near infinite energy of our universe was created from nothing, or more precisely, from random energy fluctuations. Thus, the Ultimate Free Lunch theory violated Conservation of Energy/Mass. An Integrated TOE satisfied Conservation of Energy/Mass because the energy/mass (10^{24} solar masses) in our precursor universe's super supermassive quark star (matter)/black hole (energy) equaled that in our universe.

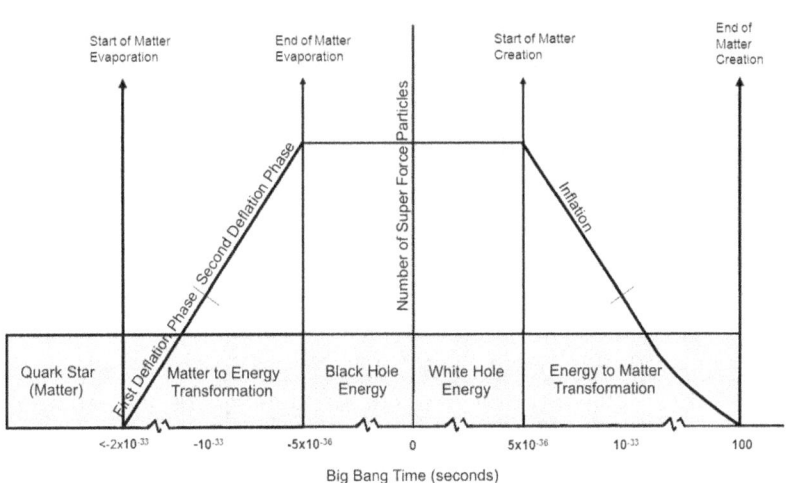

FIG. 6. Quark star/black hole to big bang (white hole) transition.

TABLE III. Ultimate Free Lunch versus an Integrated Theory of Everything.

Law	The Ultimate Free Lunch Theory	An Integrated TOE
Conservation of Energy/Mass	violates	satisfies
Einstein's Theory of General Relativity	violates	satisfies
Second Law of Thermodynamics	satisfies	satisfies

Einstein's Theory of General Relativity is time symmetrical about t = 0 and consists of a black hole, a white hole, and a wormhole connecting two universes. The Ultimate Free Lunch theory violated Einstein's Theory of General Relativity because nothing preceded our universe. In contrast, an Integrated TOE included a black hole, a white hole, and a wormhole or a doughnut shaped super force singularity in a Planck cube.

The Ultimate Free Lunch satisfied the Second Law of Thermodynamics because it assumed primacy of the latter over the laws of Conservation of Energy/Mass and Einstein's Theory of General Relativity. The logic was if our universe's entropy was a minimum at time t = 0, nothing could have preceded the big bang. Thus, the task was to prove an Integrated TOE complied with the Second Law of Thermodynamics without violating the laws of Conservation of Energy/Mass and Einstein's Theory of Relativity.

In our precursor universe, a super supermassive quark star (matter)/black hole (energy) had two time sequential states; quark star (matter) and black hole (energy). During the super supermassive quark star (matter) to black hole (energy) collapse, the maximum entropy quark star (matter) state evaporated and deflated to the minimum entropy black hole (energy) state. In a subset volume of our precursor universe, the super supermassive quark star (matter) to black hole (energy) collapse reset entropy from maximum to minimum and "resurrected" life via creation of super force or mother particles.

Big Bang Detection via Gravity Waves

The estimated big bang gravitational waveform consists of a pulse and decaying step function, both having equal maximum amplitudes. This waveform should be detectable at the big bang's location and time by an advanced extraordinarily high frequency gravitational observatory.

As shown in Fig. 7, the estimated big bang gravitational energy waveform consisted of a pulse and decaying step function. Time symmetry existed between -10^{-33} and 10^{-33} seconds because the super supermassive quark star (matter) composition at -10^{-33} seconds was identical to the hot quark-gluon plasma at 10^{-33} seconds. Gravitational energy was a maximum at -10^{-33} and 10^{-33} seconds. Between $t = 0$ and $t = 5 \times 10^{-36}$ seconds, gravitational energy was zero because matter particles had not been created. Super force particles began condensing into matter particles and their associated Higgs forces during inflation (5×10^{-36} to 10^{-33} seconds), or during the white hole (energy) to hot quark-gluon plasma (matter) transformation. At the start of the hot quark-gluon plasma (10^{-33} seconds), the heaviest matter particles were in the most compact sphere with a radius of 8 meters (see Fig. 3) and gravitational energy was a maximum. As our universe expanded following 10^{-33} seconds, matter particles moved further apart from each other and gravitational energy decreased. Thus, matter density and gravitational energy were a maximum at 10^{-33} seconds and at the corresponding hot quark-gluon plasma of the super supermassive quark star (matter) at time -10^{-33} seconds.

Prior to the deflation start time at $< -2 \times 10^{-33}$ seconds, the super supermassive quark star (matter) steadily added energy/mass and its gravitational energy increased. At the first deflation phase start time, our universe's energy/mass was spread over an extremely large (radius $\ll 10^{26}$ meters) super supermassive quark star (matter) at near zero temperature (cold quark-gluon plasma).[154] During the first deflation phase between < -2×10^{-33} and -10^{-33} seconds, the super supermassive quark star (matter) at near zero temperature collapsed to a compact hot quark-gluon plasma with a corresponding increase in gravitational energy. Lighter matter particles and their associated Higgs forces evaporated to the super force which then condensed to heavier matter particles and their associated Higgs forces. Since matter particles were further apart at the start of the first deflation phase than at the end, its gravitational energy was less. Matter evaporation during the second deflation phase was the reverse of matter creation during inflation. That is, heavier matter particles and their associated Higgs forces evaporated to super force particles between -10^{-33} and -5×10^{-36} seconds with a decrease in gravitational energy to zero at t = -5×10^{-36} seconds. Between -5×10^{-36} and 5×10^{-36} seconds, all our universe's energy (10^{54} kilograms) was in the form of super force particles and no matter particles or gravitational energy existed. That time period was also the transient life time (approximately 10^{-35} seconds) of the super supermassive black hole (energy)/white hole (energy).

The location of the estimated big bang gravitational waveform was the origin (x_u = 0, y_u = 0, z_u = 0, t = 0) of our universe's big bang, see Fig. 1. The estimated gravitational energy waveform occurred at the big bang time t = 0, or 13.7 billion years ago. If all our universe's galaxy positions are extrapolated backwards in three dimensional space, they intersect at the origin. The estimated gravitational energy waveform should be detectable at the big bang's location and time by an advanced extraordinarily high frequency (> 10^{33} Hertz) Laser Interferometer Gravitational Observatory (LIGO) or Laser Interferometer Space Antenna (LISA).

Appendix A An Integrated Theory of Everything Research Article

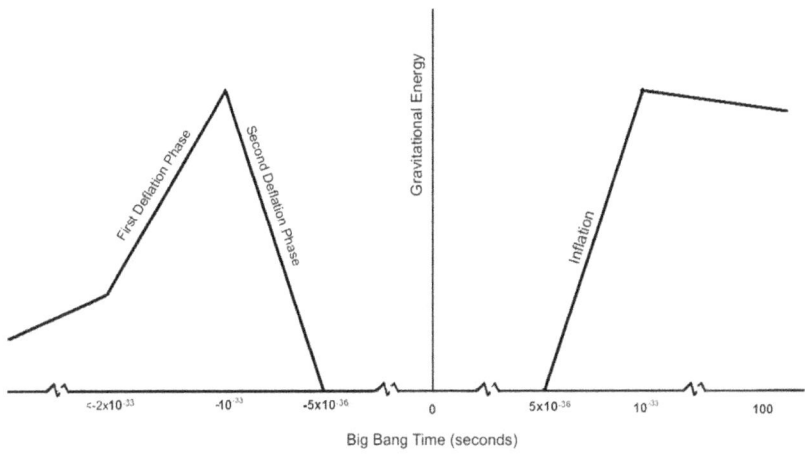

FIG. 7. Estimated big bang gravitational energy waveform.

Cosmological constant problem/Nested universes

Our universe was nested in our precursor universe which was nested in the Super Universe. The cosmological constant problem existed because the Super Universe's volume was 10^{120} larger than our universe's volume. Hubble's law existed for precursor universes within the Super Universe, universes within precursor universes, and galaxies within universes.

The observed cosmological constant was 10^{-120} of the expected value (2×10^{110} erg/cm³) and known as the cosmological constant problem.[155] According to Steinhardt, this problem existed because our universe was older than expected because of precursor cyclical universes.[156] Cyclical universes were amplified to nested universes. Cyclical universes were special cases of nested universes where the super supermassive quark star (matter)/black hole (energy) subset volume equaled the total precursor universe volume.

Fig. 8 shows three nested universes consisting of the Super Universe, our precursor universe, and our universe at four sequential big bang times (in two instead of three dimensions and not to scale). The Super Universe's big bang occurred at -10^{-50} years.[157] At an assumed t = -15 billion years, a super super supermassive black hole (energy) existed in the Super Universe which was preceded by its associated super super supermassive quark star (matter).[158] By t = 0, that super super supermassive black hole (energy) expanded into our precursor universe. Within our precursor universe, a super supermassive black hole (energy) formed preceded by its associated super supermassive quark star (matter). The super supermassive black hole (energy) transitioned to our big bang's white hole and after 13.7 billion years of expansion, our present universe exists. Fig. 8 also shows our precursor universe spawning a parallel universe at a time prior to t = 0. Within

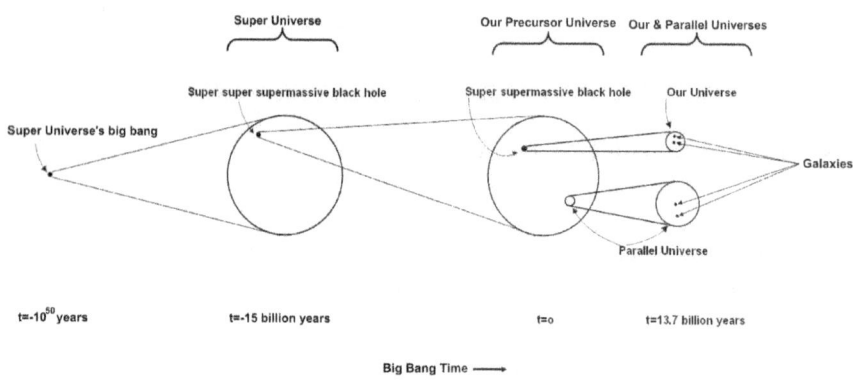

FIG. 8. Super Universe and nested universes.

our universe and the parallel universe were galaxies. Super force string singularities at the center of Planck cubes existed at the start of the Super Universe, all precursor universes, and all universes including our universe.

Fig. 9 shows three nested universes at t = 0. Our universe and a parallel universe were nested within our precursor universe. Our precursor universe was nested within the Super Universe. Dark energy density was uniformly distributed throughout the Super Universe, our precursor universe, and our universe. As the Super Universe expanded via eternal inflation, dark energy density decreased with time. Since matter was not uniformly distributed on a small scale in our precursor universe, subset volumes formed super supermassive quark stars (matter)/black holes (energy) which transitioned to white holes (e.g. our universe).[159]

The cosmological constant problem existed because the Super Universe's volume was 10^{120} larger than our universe's volume. Since spherical volumes were proportional to their radii cubed, the ratio of the Super Universe's radius R_{su} to our universe's radius R_{ou} (46.5 x 10^9 light years) was $(10^{120})^{1/3}$ or 10^{40}. The Super Universe's radius was R_{su} = (10^{40}) (46.5 x 10^9 light years) or approximately 10^{50} light years. Assuming equal expansion rates, that is, our universe's radius/our universe's age = Super Universe's radius/Super Universe's age, the Super Universe's age was approximately 10^{50} years.

Since the Super Universe's volume was 10^{120} larger than our universe, there were approximately 10^{120} parallel universes in the Super Universe. Galaxies of these parallel universes were uniformly distributed in the Super Universe between our universe's boundary (our universe's radius of 46.5 billion light years plus the unknown thickness of the spherical shell "perfect" vacuum) and the spherical Super Universe's boundary (radius of 10^{50} light years).

Hubble's law exists for precursor universes within the Super Universe, universes within our precursor universe, and galaxies within our universe as shown in Fig. 10. At the Super Universe's big bang 10^{50} years ago, all the Super Universe's energy $(10^{54}$ kilograms$)(10^{120})$ = 10^{174} kilograms was in the Super Universe's super force singularity. Precursor universes within the Super Universe were created by precursor universes' big bangs. There was a Hubble's law or a linear relationship between the velocity or red shift of these precursor universes and time or distance. Similarly, there was a Hubble's law for universes within our precursor universe.

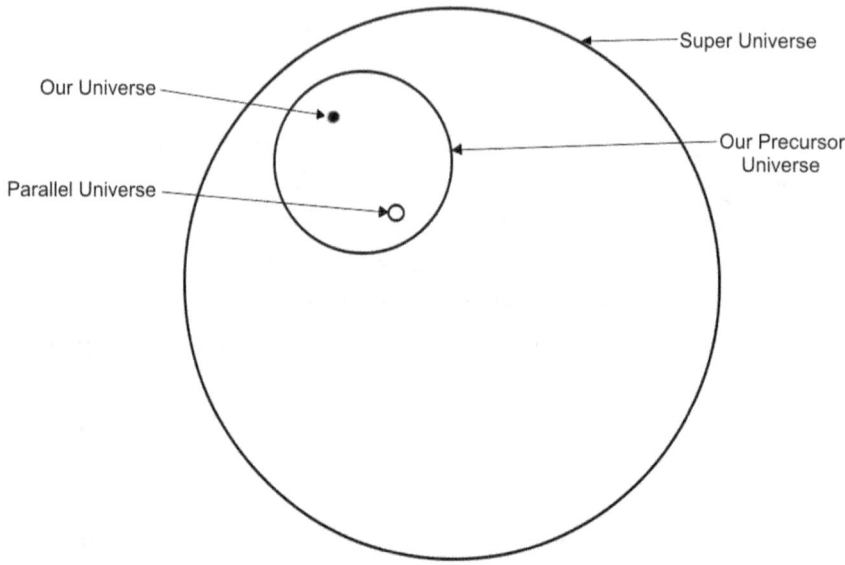

FIG. 9. Three nested universes at t = 0.

Our universe was created 13.7 billion years ago by a doughnut shaped super force singularity surrounded by a spherical "perfect" vacuum. As shown in Fig. 10, our universe decelerated for its first eight billion years and accelerated during the next 6 billion years. Currently, a spherical shell "perfect" vacuum exists between our universe and the inner boundary of our precursor universe. As our universe accelerates, the spherical shell thickness will approach zero. Our universe's acceleration will stop when our universe's boundary merges with our precursor universe's inner boundary. Eventually, the expansion rate of galaxies within our universe will become identical to the expansion rates of universes within our precursor universe and precursor universes within the Super Universe. This is shown by three equal slopes at a time greater than 13.7 billion years.[160]

The Hubble Ultra Deep Field telescope can detect galaxies with an age of 13.1 billion years. The James Webb telescope will detect Population III stars and galaxies several hundred million years older. An advanced telescope is required to detect the closest galaxy in the closest parallel

Appendix A An Integrated Theory of Everything Research Article

universe of our precursor universe, that is, a galaxy with an age greater than 13.7 billion years. This integrates the cosmological constant problem with Super Universe, dark energy, stellar black holes, black hole entropy, and arrow of time theories, (see Table IV).

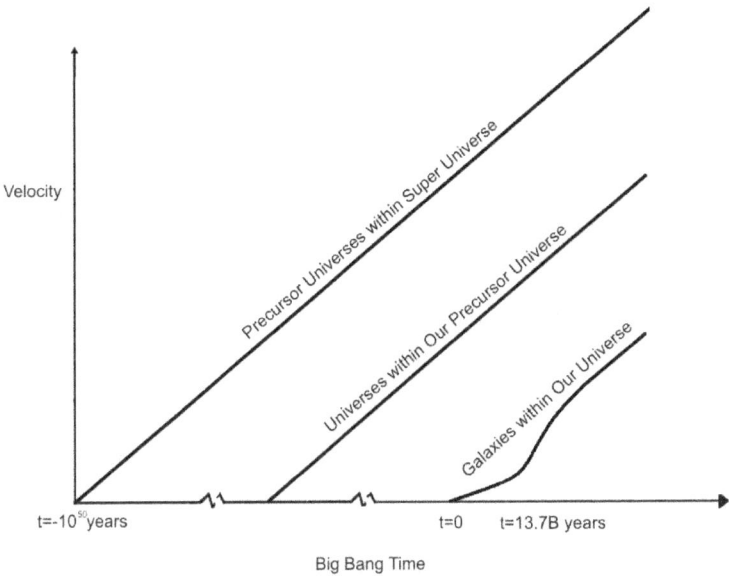

FIG. 10. Hubble s law.

Black hole information paradox

Any universe object's intrinsic information consists of the contents and positions of all the object's contiguous Planck cubes. Intrinsic information is lost in a super supermassive quark star (matter)/black hole (energy) formation and none is emitted as Hawking radiation.

In 1975, Hawking stated Hawking radiation contained no information swallowed by a black hole. In 2004, his position reversed and Hawking radiation contained information. This is the black hole information paradox.

The "No Hair" theorem states a stellar black hole (energy) has three information parameters; mass, charge and spin, whereas our universe contains near infinite information. Any universe object's (e.g. an encyclopedia) intrinsic information at a time = t consists of the contents and positions (x_u, y_u, z_u, t of Fig. 1) of all the object's contiguous Planck cubes. Intrinsic information consists primarily of the unique relative orientation of up quarks, down quarks, and electrons to each other, or an object's molecular, atomic, nuclear, and fundamental matter (e.g. up quark) structure. In contrast, a universe object's (e.g. an encyclopedia) extrinsic information consists of its written words. An encyclopedia and a pile of manure having the encyclopedia's identical dimensions and number of Planck cubes have comparable but different intrinsic information. In contrast, the encyclopedia has significant extrinsic information (e.g. its written words) whereas the identical pile of manure has none.

Each up quark, down quark, and electron resides within a specific Planck cube of the encyclopedia's ink, paper, binding, etc. molecules. Encyclopedia intrinsic information is lost in four star stages during decomposition of its molecules to atoms, to protons/neutrons and electrons, to quarks, and to super force energy. In a white dwarf star, molecules decompose to atoms. In a neutron star, atoms decompose to neutrons, protons, and electrons. In a super supermassive quark star (matter), protons and neutrons decompose to up and down quarks. In a super supermassive black hole (energy), up and down quarks decompose (evaporate) to super force particles. Intrinsic or structural information is lost in a super supermassive quark star (matter)/black hole (energy) formation and none is emitted as Hawking radiation.

This integrates black hole information paradox with stellar black holes and particle creation theories (see Table IV).

Baryogenesis

Charge, parity, and time (CPT) violation was the theory which best explained baryogenesis. There were three CPT violation arguments which supported each other and conclusions of previous sections. CPT, unitarity, and entropy preservation were violated in the highly curved space-times of both our precursor universe's super supermassive black hole (energy) and its big bang white hole (energy) counterpart.

Baryogenesis is the asymmetric production of baryons and anti-baryons in the early universe expressed as the baryon to photon ratio η = 6.1 x 10^{-10}.[161] There are 42 identified baryogenesis theories of which six are prominent; electroweak, GUT, quantum gravity, leptogenesis, Affleck-Dine, and CPT violation.[162] Electroweak occurs insufficiently in the Standard Model and is considered unlikely without supersymmetry. Inflationary scenarios disfavor GUT and quantum gravity theories. Leptogenesis and Affleck-Dine are viable but not well understood.[163]

The sixth baryogenesis theory is CPT violation which has three mutually supportive arguments. The first argument is the CPT theorem is invalid at the Planck scale.[164]

According to the CPT theorem, laws of physics are unchanged by combined CPT operations provided locality, unitarity (sum of all possible outcomes of any event is one), and Lorentz invariance are respected. In the second argument, highly curved space-times such as a super supermassive black hole (energy) singularity violate CPT because of apparent violations of unitarity caused by incoming matter information disappearance.[165] From the black hole information paradox section's conclusion, incoming matter information is lost in a super supermassive black hole (energy) formation.

The third argument is a quantum mechanics axiom states the evolution of a system, or the transformation from one state to another, must be unitary. Entropy is preserved in unitary dynamics.[166] In a super supermassive quark star (matter) to black hole (energy) collapse, energy/mass quanta in Planck cubes collapse to a super force singularity (no quanta). Thus, quantum mechanics is invalid and unitarity and entropy preservation are violated. From the arrow of time section's conclusion, in a super supermassive quark star (matter) to black hole (energy) collapse, entropy switches from maximum to minimum so entropy is not preserved.

CPT, unitarity, and entropy preservation were violated in the highly curved space-times of both our precursor universe's super supermassive black hole (energy) and its big bang white hole (energy) counterpart. Each matter particle's transformation to a super force particle and each super force to matter particle transformation violated CPT, which provided sufficient CPT violations to produce our universe's baryon to photon ratio of 6.1×10^{-10}.

This integrates baryogenesis with black hole information paradox, arrow of time, stellar black holes, and black hole entropy theories (see Table IV).

Quantum Gravity Theory

Quantum gravity, string theory, and an Integrated TOE unify all known physical phenomena from the infinitely small or Planck cube scale (quantum mechanics) to the infinitely large or Super Universe scale (Einstein's General Relativity). Quantum gravity theories include string theory.

Quantum gravity is an evolving theory that unifies quantum mechanics at the infinitely small Planck cube scale to Einstein's General Relativity at the infinitely large Super Universe scale. All matter and force particles exist as strings and reside within our universe's fundamental building block, the Planck cube. Since the Planck cube is the quantum or unit of matter particles, force particles, and space, its actions are described by quantum mechanics. Extremely massive and dense bodies such as collapsed stars of the infinitely large Super Universe are governed by Einstein's law of General Relativity. Collapsed stars include; white dwarfs, neutron stars, supermassive quark stars (matter), super supermassive quark stars (matter)/black holes (energy), and super super supermassive quark stars (matter)/black holes (energy).

Appendix A An Integrated Theory of Everything Research Article

String theory defined all fundamental matter and force particles as strings in Planck cubes. Any object in the Super Universe can be represented by a volume of contiguous Planck cubes containing fundamental matter or force particle strings. Super force string singularities at the center of Planck cubes existed at the start of the Super Universe, all precursor universes, and all universes including our universe. Thus, string theory unified quantum mechanics of the infinitely small at the Planck cube scale (e.g. fundamental matter and force particles) with Einstein's General Relativity of the infinitely large at the Super Universe scale (e.g. the super super super supermassive black hole (energy) or doughnut shaped super force singularity which created the Super Universe).

Quantum gravity, string theory, and an Integrated TOE unified all known physical phenomena from the infinitely small or Planck cube scale (quantum mechanics) to the infinitely large or Super Universe scale (Einstein's General Relativity). This integrated quantum gravity with all other nineteen theories in an Integrated Theory of Everything (see Table IV).

Conclusions

An Integrated TOE unified all known physical phenomena from the infinitely small or Planck cube scale to the infinitely large or Super Universe scale. Each matter and force particle existed within the universe's fundamental building block, the Planck cube. Any universe object was representable by a volume of contiguous Planck cubes. The Planck cube was the quantum or unit of matter particle, force particle, and space. An Integrated TOE unified 16 Standard Model particles, 16 supersymmetric particles, 32 anti-particles, their 64 associated supersymmetric Higgs particles, and the super force or mother particle for 129 particles.

Each of 129 fundamental matter and force particles was represented by its unique string or associated Calabi-Yau membrane in a Planck cube. A string or associated Calabi-Yau membrane's energy/mass was primarily a function of its diameter and secondarily its hill's and valley's amplitude displacement and frequency. The big bang's near zero diameter singularity of superimposed super force strings consisted of our universe's near infinite energy. Any object in the Super Universe could be represented by a volume of contiguous Planck cubes containing fundamental matter or force particle strings. Super force string singularities at the center of Planck cubes existed at the start of the Super Universe, all precursor universes, and all universes including our universe.

Two reasons for replacing inadequate existing symbols with proposed symbols were; explicit and amplified Higgs particle representation and elimination of existing symbol ambiguities via standardization of subscripts and capitals.

The big bang created our universe's 128 particles from the super force having energy of 10^{54} kilograms. Matter creation was time synchronous with both the inflationary period start time and the one to seven Planck cubes energy to matter expansion. By t = 100 seconds, all super force energy had condensed into eight permanent matter particles and their eight associated Higgs force energies.

The process of generating 17 matter particles and their 17 associated Higgs forces was spontaneous symmetry breaking or the Higgs mechanism. The sum of eight permanent Higgs forces' energies associated with eight permanent matter particles: atomic matter (up quark, down quark, electron); dark matter (zino, photino); and neutrino matter (tau-neutrino, muon-neutrino, electron-neutrino) constituted dark or vacuum energy.

The 32 standard and supersymmetric matter and force particles and their 32 anti-particles were supersymmetric with 64 associated Higgs particles and the latter were supersymmetric with themselves. There were three types of spontaneous symmetry breaking functions for three types of matter particles: 17 standard and supersymmetric matter particles, 3 Standard Model Higgsinos, and 12 supersymmetric Higgsinos.

Intermediate force particles were W/Z bosons for Standard Model particles and supersymmetric W/Z_{ss} bosons for supersymmetric particles.

Appendix A An Integrated Theory of Everything Research Article

Decays were a series of evaporations of matter particles and their associated Higgs forces to the super force and condensations from the super force to less massive matter particles and their associated Higgs forces. The neutral heavy lepton was a constituent of dark matter.

Dark matter consisted of zinos and photinos. Dark matter agglomeration formed the framework of galaxies.

There were four sequential universe expansions. Entropy increase of the super force and its derivatives drove the expansion within our universe's first Planck cube. X bosons' (inflatons) latent heat drove the inflationary period's exponential expansion. Dark energy drove both the uniform and non-uniform distribution of matter expansions. String theory's seventh extra dimension was the product of our universe's non-uniform distribution of matter expansion rate and the graviton's intergalactic propagation time.

By the end of matter creation or $t = 100$ seconds, our universe consisted of baryonic matter (4.6 %), cold dark matter (22.8%), and dark energy (72.6%), and these percentages remained constant for 13.7 billion years. The cosmological constant was proportional to vacuum or dark energy density. Dark energy density was the sum of eight permanent Higgs force densities.

Messenger particles were amplified with embedded clock/computers as their operational mechanisms. The graviton/photon clock/computer calculated Newton's gravitational or Coulomb's force and provided it to the receiving particle. The gluon clock/computer calculated the strong force and provided it to the receiving quark. The relative strengths of gravitational and electromagnetic/weak forces were due to propagation factor dilution between gravitational force activation and electromagnetic/weak force creation/activation.

A stellar black hole was a quark star (matter) or black hole (energy) both of which were "black." Six types of stellar black holes were: supermassive quark star (matter), quark star (matter), super supermassive quark star (matter), its associated super supermassive black hole (energy), super super supermassive quark star (matter), and its associated super super supermassive black hole (energy). Our precursor universe's super supermassive quark star (matter)/black hole (energy) created our universe's "big bang" (white hole) via conservation of energy/mass.

The proposed entropy formula for a quark star (matter) was proportional to the quark star's volume (r^3) and inversely proportional to a Planck cube's volume $(l_p)^3$.

In our universe and our precursor universe, entropy increased with time. Our universe was created by a doughnut shaped super force singularity of a super supermassive black hole (energy), surrounded by a spherical "perfect" vacuum. Our precursor universe's maximum entropy super supermassive quark star (matter) evaporated, deflated, and collapsed to the minimum entropy black hole (energy), "resurrecting" life.

An Integrated TOE satisfied Conservation of Energy/Mass, Einstein's Theory of General Relativity, and the Second Law of Thermodynamics for our precursor universe's super supermassive quark star (matter)/black hole (energy) transition to the big bang.

The estimated big bang gravitational waveform consisted of a pulse and decaying step function, both having equal maximum amplitudes. This waveform should be detectable at the big bang's location and time by an advanced extraordinarily high frequency gravitational observatory.

Our universe was nested in our precursor universe which was nested in the Super Universe. The cosmological constant problem existed because the Super Universe's volume was 10^{120} larger than our universe's volume. Hubble's law existed for precursor universes within the Super Universe, universes within precursor universes, and galaxies within universes.

Appendix A An Integrated Theory of Everything Research Article

Any universe object's intrinsic information consisted of the contents and positions of all the object's contiguous Planck cubes. Intrinsic information was lost in a super supermassive quark star (matter)/black hole (energy) formation and none was emitted as Hawking radiation.

Charge, parity, and time violation was the theory which best explained baryogenesis. There were three CPT violation arguments which supported each other and conclusions of previous sections. CPT, unitarity, and entropy preservation were violated in the highly curved space-times of both our precursor universe's super supermassive black hole (energy) and its big bang white hole (energy) counterpart.

Quantum gravity, string theory, and an Integrated TOE unified all known physical phenomena from the infinitely small or Planck cube scale (quantum mechanics) to the infinitely large or Super Universe scale (Einstein's General Relativity). Quantum gravity theories included string theory.

Twenty independent existing theories were replaced by an Integrated TOE consisting of twenty interrelated amplified theories. Table IV Primary interrelationships between twenty amplified theories summarized the interrelationships of an Integrated TOE.

Table IV. Primary interrelationships between twenty amplified theories.

	String	Particle creation	Inflation	Spontaneous symmetry breaking	Higgs forces/supersymmetric Higgs particles	Superpartner and quark decays	Neutrino oscillations	Dark matter	Universe expansions	Dark energy	Messenger particles	Relative strengths of forces	Super Universe	Stellar black holes	Black hole entropy	Arrow of time	Cosmological constant problem	Black hole information paradox	Baryogenesis	Quantum gravity
String	x	x												x						x
Particle creation	x	x	x	x	x		x	x	x		x		x					x	x	x
Inflation		x	x	x	x			x	x	x									x	x
Spontaneous symmetry breaking		x	x	x	x	x	x	x		x									x	x
Higgs forces/supersymmetric Higgs particles		x	x	x	x			x	x	x	x								x	x
Superpartner and quark decays				x		x														x
Neutrino oscillations			x			x	x													x
Dark matter		x	x	x	x		x	x			x								x	x
Universe expansions		x	x		x				x			x								x
Dark energy		x	x	x	x			x		x			x	x	x	x	x		x	x
Messenger particles				x							x		x							x
Relative strengths of forces		x							x			x								x
Super Universe									x				x	x	x	x	x			x
Stellar black holes	x	x							x				x	x	x	x	x	x	x	x
Black hole entropy									x				x	x	x	x	x	x	x	x
Arrow of time									x				x	x	x	x	x	x	x	x
Cosmological constant problem									x				x	x	x	x	x			x
Black hole information paradox		x											x	x	x				x	x
Baryogenesis		x	x	x	x			x		x				x	x	x			x	x
Quantum gravity	x	x	x	x	x	x	x	x	x	x	x	x	x	x	x	x	x	x	x	x

Research Article Endnotes

[102.] The Planck cube quantum was selected for two reasons, Planck units and string theory. Planck units consist of the following five normalized, natural, universal, physical constants; gravitational constant, reduced Planck constant, speed of light in a vacuum, Coulomb constant, and Boltzmann constant. The Planck length which defines a Planck cube is a function of three of the five constants; gravitational constant, reduced Planck constant, and the speed of light in a vacuum. In string theory, the Planck length is the size of matter and force particle strings.

[103.] Because of three star factor products only a small portion of our universe's volume (estimated at 10-50) contained stellar black holes. The three factor products were: stars were concentrated matter surrounded by large volumes of space (10-30); only a small fraction of stars were stellar black holes (10-3); and stellar black holes were compressed stars (10-17). Thus, Newton's equations of motion and Cartesian Coordinates were applicable for most of our universe's volume. In those sub-volumes containing stellar black holes and on an exception basis, Einstein's equations of General Relativity were substituted for Newton's equations and space-time coordinates substituted for Cartesian Coordinates.

In the future, the universal rectangular coordinate system (Cartesian Coordinates) should also originate at the "Super Universe's" big bang.

[104.] B. Greene, *The Elegant Universe* (Vintage Books, New York, 2000), p. 144. Greene specifies only amplitude displacement and wavelength (frequency). He also describes Calabi-Yau membranes as beach balls, doughnuts, and multi-doughnuts, and conifold transitions as the procedure whereby membranes transition into each other, pp. 327-329. Energy (E) is inversely proportional to diameter (d), for example, $E = 1/d^n$, where n is an exponent.

[105.] G. W. Hinshaw, http://map.gsfc.nasa.gov/universe/uni_matter.html. (2010). Click on Our Universe (Matter/Energy). Mass = (Universe volume) (density) = $[4\pi(4.4 \times 10^{26} \text{ meters})^3/3] [9.9 \times 10^{-27} \text{ kilograms/meters}^3] = 3.5 \times 10^{54}$ kilograms. Near infinite is defined as finite but extremely large.

[106.] The relationship between quantum numbers and particle location should be analyzed. For example, the relationship between the four quantum numbers of an electron in an atom and the electron's location should be extended to "free" fundamental particles such as electrons and up quarks in a quark-gluon plasma.

[107.] The W/Z bosons (p_{15}) are actually transient matter particles with associated Higgs force particles (h_{15}) instead of force particles (bosons) with associated Higgsino matter particles.

[108.] M. Rees, Ed., *Universe*. (DK Publishing, New York, 2005), pp. 46-49.

[109.] The W/Z bosons were represented by one p_{15} instead of three (W^+, W^-, and Z^0 or p_{15a}, p_{15b}, and p_{15c}) particles.

[110.] A. H. Guth, *The Inflationary Universe* (Perseus Publishing, New York, 1997), p. 185 Fig. 10.6. Fig. 3 initialized the inflationary period start radius at $.8 \times 10^{-35}$ meters with an exponential inflation factor of 10^{36}. Guth's comparable values were 10^{-50} meters and 10^{49}. Liddle and Lyth specify an exponential inflation factor greater than 10^{26}. (*Cosmological Inflation and Large-scale Structure*, p 46)

[111.] The end of matter creation was assumed to be the end of the lightest anti-matter particle or the anti-electron-neutrino. Anti-electron-neutrinos existed after 100 seconds. However, since the end time of anti-electron-neutrinos was unknown, the end of matter creation was approximated to be 100 seconds or the end of anti-electrons.

Research Article Endnotes

[112.] W/Z bosons are transient matter particles which cause an asymmetrical number of matter particles (i.e. 17 instead of 16).

[113.] Twelve superpartners were assumed to exist at < 10^{-36} seconds because they were the latent heat source during inflation.

[114.] A. H. Guth, *The Inflationary Universe* (Perseus Publishing, New York, 1997), p. 209 Fig. 12.1, pp. 140-3.

[115.] During baryogenesis, the ball initially at its peak position $h_{11} = h_{11bar} = 0, Z = 2$, moves down the spontaneous symmetry breaking function equidistant between the X and Y axes. Super force particles condense into a particle p_{11}, its associated Higgs force h_{11}, an anti-particle p_{11bar}, and its associated Higgs force h_{11bar}. The four particles then annihilate and evaporate back to super force energy as the ball returns to its initial peak position (bidirectional spontaneous symmetry breaking). During the second cycle, the ball moves down the spontaneous symmetry breaking function closer to the X axis and then back to its original position. After n condensation/evaporation cycles in the false vacuum state, the ball eventually moves to the Fig. 4 ball position ($h_{11} = -2$, $h_{11bar} = 0$, $Z = 1.5$) or the true vacuum state where the super force condenses totally to the permanent up quark p_{11} and its associated Higgs force h_{11}.

[116.] The Higgs force is energy and not a point particle such as an up quark. Therefore, the Large Hadron Collider should search for Higgs force energy not Higgs force point particles having cross sections.

[117.] A second type of bidirectional spontaneous symmetry breaking is subsequently described for the collapse of a super supermassive quark star (matter) to a super supermassive black hole (energy) where permanent matter particles and their associated Higgs forces evaporate back to super force particles.

[118.] B. Povh, K. Rith, C. Scholz, and F. Zetsche, *Particles and Nuclei* (Springer-Verlag Berlin, Heidelberg, 2008), p. 2.

[119.] However, density calculations if used must be modified.

120. Although 17 types of super force particles are described in this article, there are actually 64 types of super force particles which experience spontaneous symmetry breaking and condense into 32 matter and force particles, 32 anti-matter and force particles, and their 64 associated Higgs particles. However, baryogenesis eliminates half of these, so there are just 32 super force types which condense into 17 standard (13) and supersymmetric (4) matter particles and their 17 associated Higgs forces and 15 Higgsinos and their 15 standard (3) and supersymmetric (12) associated forces.

121. Since the stable LSPs (zinos and photinos) were assumed formed by 10^{-12} seconds or an approximate temperature of 10^{15} K, supersymmetric W/Z_{ss} bosons do not exist after that time or temperature. Therefore, there can be no indirect detection of dark matter (zinos and photinos) via annihilation products in our universe's galaxies which are at lower temperatures than 10^{15} K.

122. G. Kane, Sci. Am. **293**, 40-48 (July 2005).

123. B. Kayser, http://www.pd.infn.it/~laveder/unbound/scuole/2009/DBD-09/B_Kayser-DBDmeeting-oct-2009.pdf. (2009). The NHL is referred to as a large right-handed Majorana.

124. M. Y. Khlopov, http://www.roma1.infn.it/people/bini/seminars/khlopov.ppt. (2006).

125. D. B. Cline, Sci. Am. **288**, 53 (March 2003). As described in the Supersymmetric Higgs particles section, 15 Higgsinos were produced during spontaneous symmetry breaking. An undefined subset of these including h_1, h_2, and h_{16} associated with the graviton p_1, gluon p_2, and photon p_{16}, should eventually be included as dark matter components.

126. M. Turner's estimate (private communication).

127. G. W. Hinshaw, http://arxiv.org/pdf/0803.0732v2.pdf (2008).

128. E. J. Chaisson, https://www.cfa.harvard.edu/~ejchaisson/cosmic_evolution/docs/text/text_part_5.html (2012).

129. Galaxies were created approximately 200 million years after the start of our universe when population III stars formed, collapsed, and created prototype galaxies.

130. e_r is itself a function of time because our universe decelerated during its first 8 billion years and accelerated during the last 6 billion years.

131. A second reason for a constant total baryonic energy/mass was rest mass was converted to kinetic energy and radiation during nucleosynthesis. For example, stellar nucleosynthesis for stars heavier than our sun occurred via the CNO (carbon-nitrogen-oxygen) cycle. In this cycle, 4 protons fused and produced an alpha particle, 2 positrons, 2 electron neutrinos, 3 gamma rays, and 26.8 MeV of energy. The energy appeared as kinetic energy of the products. Also, the 3 gamma rays were eventually absorbed by matter particles and converted into kinetic energy. Thus, the total baryonic energy/mass remained constant at 4.6%.

132. S. M. Carroll, http://preposterousuniverse.com/writings/encyc/. (2000).

133. A. Zichichi, http://cerncourier.com/cws/article/cern/38704. (2009).

134. Upon creation of both the force carrier photon (p16b) and the electromagnetic radiation photon (p16a) during condensation of the super force to the Higgsino matter particle (h16) and its associated photon (p16) via spontaneous symmetry breaking, both photon types had zero energy. During the force carrier photon (p16b) transmission described in this section, the force carrier photon (p16b) also had zero energy. The handedness characteristic, discrimination among multiple fermion forces, and messenger particle repetition rates should be analyzed and incorporated into an updated force carrier messenger particle theory.

In contrast, during electromagnetic radiation photon (p16a) transmission (e.g. the sun's infrared transmission to earth), the electromagnetic radiation messenger particle contained energy proportional to radiation frequency or E = hf where E is energy, h is Planck's constant, and f is radiation frequency.

[135.] M.A. Thomson, http://www.hep.phy.cam.ac.uk/~thomson/lectures/partII-particles/pp2004_qcd.pdf. (2004).

[136.] C. P. Poole, *The Physics Handbook*, (John Wiley, New York, 1998), p. 365 Fig. 25-1. Electromagnetic and weak force strengths are equal under 10^{-18} meters.

[137.] Smaller types of black holes (e.g. micro) were not analyzed.

[138.] C. Miller (private communication). The discrepancy between the initial star mass (8 solar) and the final mass (1.38 solar or Chandrasekhar limit) is due to winds.

[139.] D. Leahy, R. Ouyed, http://arxiv.org/PS_cache/arxiv/pdf/0708/0708.1787v4.pdf. (2008). The quark star (matter) formed following a quark-nova's confinement energy release. The delayed secondary explosion followed a neutron star's primary supernova explosion.

[140.] D. Savage, http://hubblesite.org/newscenter/archive/releases/1997/01/text/. (1997).

[141.] C. Carilli, Science **323**, 323 (2009). Galaxy to central black hole mass ratio was 30:1 in the early universe and 700:1 now.

[142.] R. Irion, Science **295**, 66 (2002).

[143.] If a supermassive quark star (matter) accreted all star/matter in its vicinity it could appear as a void.

[144.] A. Kurkela, P. Romatschke, A. Vuorinen, http://arxiv.org/PS_cache/arxiv/pdf/0912/0912.1856v2.pdf. (2010).

145. J. D. Bekenstein, http://arxiv.org/PS_cache/quant-ph/pdf/0311/0311049v1.pdf. (2003).

146. A. Dabholkar, Current Science **89**, 2059 (2005).

147. A. Hamilton, http://casa.colorado.edu/~ajsh/schww.html (2001). A white hole is the reverse of a black hole. A black hole swallows matter and energy whereas a white hole expels it.

148. B. Greene, *The Fabric of the Cosmos* (Vintage Books, New York, 2005), p. 173.

149. K. Griest, http://physics.ucsd.edu/students/courses/winter2010/physics161/p161.3mar10.pdf. (2010).

150. Permanent matter particles (e.g. up quark) are represented by step functions having finite rise times and centered at each matter particle's condensation time. A transient matter particle (e.g. top quark) is represented by, for example, a Gaussian function centered at its condensation time and with a finite standard deviation.

151. At $t = 10^{-33}$ seconds or the quark era start, our universe consisted of a hot quark-gluon plasma.

152. The super supermassive quark star (matter) was a sphere consisting of a near zero temperature quark-gluon plasma with eight permanent matter particles: atomic matter (down quark, up quark, and electron); neutrino matter (tau-neutrino, muon-neutrino, and electron-neutrino) and dark matter (photino, zino).

The super supermassive black hole (energy) was a rotating, charged, doughnut shaped super force singularity at the center of a Planck cube. This singularity was also known as a Kerr-Newman black hole.

153. The term "Ultimate Free Lunch" is attributed to Dr. Alan Guth based on a paper by Edward Tryon, *"Is the Universe a Vacuum Fluctuation,"* Nature **246**, 396-7 (1973). See A. H. Guth, *The Inflationary Universe* (Perseus Publishing, New York, 1997), chapters 1 and 17.

154. The radius of the extremely large super supermassive quark star (matter) which created our universe was much less than 10^{26} meters, or $\ll 10^{26}$ meters, and estimated as follows. The Schwarzschild radius which defined the event horizon for a non-rotating quark star was 4×10^{26} meters for our universe's mass, or $R_s = (2G/c^2)$ (m) = $(1.48 \times 10^{-27}$ m/kg) $(.274 \times 10^{54}$ kg) = 4×10^{26} m, where R_s is the Schwarzschild radius, G is the gravitational constant, c is the velocity of light, and m is mass. This was the upper radius limit. Since the super supermassive quark star's (matter) equation of state and its cold quark-gluon plasma density were unknown, the lower radius limit was estimated as follows. The theoretical lower radius limit occurred when all matter particles of the super supermassive quark star (matter) were in contiguous Planck cubes. If each matter particle existed in a Planck cube and there were 10^{81} matter particles, the super supermassive quark star's (matter) volume was V = $(1.6 \times 10^{-35}$ m$)^3$/ (matter particle) (10^{81} matter particles) = 4×10^{-24} m^3 or a radius of approximately 10^{-8} meters. The estimated super supermassive quark star's (matter) radius was between the upper (4×10^{26} meters) and lower (10^{-8} meters) radius limits or approximately $\ll 10^{26}$ meters.

The start of deflation or matter evaporation in Fig. 6 and Fig. 7, was estimated at less than twice the inflation time ($< -2 \times 10^{-33}$ seconds) as follows. During the inflation time of approximately 10^{-33} seconds, our universe expanded in size from a radius of $.8 \times 10^{-35}$ meters to a radius of 8 meters for an exponential inflation factor of 10^{36}. To achieve a radius of 4×10^{26} meters from a radius of 8 meters requires an additional exponential inflation factor of $4 \times 10^{26}/8 \sim 10^{26}$ which is less than 10^{36}. Therefore, assuming identical exponential inflation/deflation rates and the upper limit radius of 4×10^{26} meters, less than twice the inflation time or $< 2 \times 10^{-33}$ seconds would be required for our precursor universe's super supermassive quark star (matter) to deflate from 4×10^{26} meters to $.8 \times 10^{-35}$ meters.

155. S. M. Carroll, http://www.livingreviews.org/lrr-2001-1. (2008).

156. P. J. Steinhardt and N. Turok, *Endless Universe: Beyond the Big Bang* (Doubleday, New York, 2007), p 249.

[157.] Eventually the big bang time scale where our universe's big bang occurred at t = 0, should be replaced by the start of the Super Universe where t = 0 occurred 10^{50} years ago.

[158.] Fig. 8 shows one precursor universe between the Super Universe and our universe. However, there could be from 0 to n time sequential precursor universes. For n = 0, the Super Universe was our precursor universe. In general, there were n nested time sequential precursor universes.

To provide a variety of sizes for a quark star (matter) to its associated black hole (energy) collapse, collapse size was assumed to be a function of two thresholds, energy/mass and energy/mass density. For creation of our universe, the energy/mass threshold was 10^{54} kilograms and the associated energy/mass density was ρ_{ou} (where ou signifies our universe) and currently undefined. If only one collapse threshold existed (e.g. energy/mass), any super supermassive quark star (matter) greater than 10^{54} kilograms would collapse to its associated super supermassive black hole (energy) before it grew any larger. A super super supermassive quark star (matter) was assumed to have an energy/mass collapse threshold much greater than 10^{54} kilograms and an energy/mass density collapse threshold different than ρ_{ou}. There were thus many combinations of energy/mass and energy/mass density thresholds for creation of a variety of super supermassive quark stars (matter)/black holes (energy) sizes in precursor universes and super super supermassive quark stars (matter)/black holes (energy) sizes in the Super Universe.

[159.] Matter is currently uniformly distributed on a large scale in our universe where large scale is defined as a cube with a side equal to approximately 300 million light years. See R. P. Kirshner, *The Extravagant Universe: Exploding Stars, Dark Energy and the Accelerating Cosmos*, (Princeton University Press, Princeton, 2002), p. 71.

[160.] The simplest Hubble's law of equal and constant expansion was assumed in Fig. 10 for the three categories; precursor universes within the Super Universe, universes within our precursor universe, and galaxies within our universe.

161. J. M. Cline, http://arxiv.org/PS_cache/hep-ph/pdf/0609/0609145v3.pdf. (2006). Big bang nucleosynthesis determined η and the Wilkinson Microwave Anisotropy Probe measured it accurately.

162. M. Shaposhnikov, http://www.physicsforums.com/archive/index.php/t-53648.html. (2004).

163. N. Bao, P. Saraswat, http://www.astro.caltech.edu/~golwala/ph135c/14SaraswatBaoBaryogenesis.pdf. (2007).

164. T. D. Lee, *Selected Papers, 1985-1996* (Gordon and Breach, Amsterdam, 1998), p 776, p 787.

165. N. E. Mavromatos, http://arxiv.org/PS_cache/hep-ph/pdf/0504/0504143v1.pdf. (2005).

166. F. Hulpke et al., Foundations of Physics **36**, 479, 494 (2006).

167. I am grateful to all my formal and informal educators.

Appendix B An Integrated Theory of Everything Video Presentation

An Integrated Theory of Everything research article in Appendix A was published on the Internet at:

http://toncolella.files.wordpress.com/2012/07/m080112.pdf

A 60 minutes video summary of that article called "An Integrated Theory of Everything Video Presentation (YouTube)" is available at:

http://www.youtube.com/watch?v=CD-QoLeVbSY

www.ingramcontent.com/pod-product-compliance
Lightning Source LLC
Chambersburg PA
CBHW061506180526
45171CB00001B/52